DON'T KNOW
MUCH ABOUT®
The Universe

D0289050

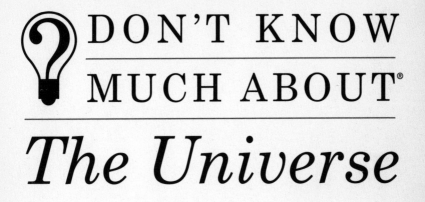

DON'T KNOW

MUCH ABOUT®

The Universe

EVERYTHING
YOU NEED TO KNOW
ABOUT OUTER SPACE
BUT NEVER LEARNED

KENNETH C. DAVIS

Perennial

An Imprint of HarperCollinsPublishers

The Library of Congress has catalogued the hardcover edition as follows:

Davis, Kenneth C.
    Don't know much about the universe: everything you need to know about the cosmos but never learned / Kenneth C. Davis.—1st ed.
        p.   cm.
    Includes bibliographical references and index.
    ISBN 0-06-019459-6
    1. Cosmology—Miscellanea. I. Title.

QB981 .D275   2001
523.1—dc21   2001024605

ISBN 0-06-093256-2 (pbk.)

02 03 04 05 06 ❖/RRD 10 9 8 7 6 5 4 3 2 1

For Star Gibbs,
who gave me my first rocket

The Way of Heaven has no favorites.
It is always with the good man.
—LAO-TZU

ACKNOWLEDGMENTS

This book, like all the books in the Don't Know Much About series, was a collaborative effort, with a great many people assisting me in my work. For their steady support, sound advice, and good cheer, I first would like to thank all my friends at the David Black Agency: David Black, Leigh Ann Eliseo, Gary Morris, Susan Raihofer, Laureen Rowland, Joy Tutela, Carmen Rey, Jason Sacher, and Doron Taleporos.

With this book, the Don't Know Much About series has found a new home and friends at HarperCollins. I would like to thank all of the many supportive people there who have welcomed me, offered assistance and encouragement, and made this book possible, including Cathy Hemming, Susan Weinberg, Carie Freimuth, Christine Caruso, Laurie Rippon, Julia Serebrinsky, Adrian Zackheim, Roberto de Vicq de Cumptich, Nora Reichard, David Koral, and Olga Gardner Galvin. I would especially like to thank my understanding, agreeable, and most helpful editor, Gail Winston, and her assistant, Christine Walsh.

My family is always my greatest source of inspiration and support. They are also very tolerant of me when writing books drags on far too long. For the good humor, love, and patience they show, I always thank my children: Jenny Davis, for her ideas and ideals; and Colin Davis, who is also my valued science and math advisor. My wife, Joann Davis, has always made my work possible. For that and all the other things she gives me, I am forever grateful to her.

CONTENTS

When I consider thy heavens, the work of thy fingers, the
   moon and the stars, which thou has ordained;
What is man, that thou art mindful of him? and the son of
   man, that thou visitest him?
For thou hast made him a little lower than the angels.

Psalms 8:23–25

Astronomy compels the soul to look upwards and leads us
from this world to another.

PLATO, The Republic

Space isn't remote at all. It's only an hour's drive away if your
car could go straight upwards.

SIR FRED HOYLE, in the London Observer, 1979

"*Scientists confirmed today that everything we know about the structure of the universe is wrongedy-wrong-wrong.*"

INTRODUCTION

LOST IN SPACE

When I was a teenager, I used to pose questions like this: "Mom, can I borrow five bucks for a movie?" Or, "Mrs. Brown, can I get an extension on that term paper?" Or, "Will I ever get a date?"

When Albert Einstein was a teenager, he asked, "What would the world look like if I rode on a beam of light?"

I don't know why Einstein's extraordinary query never occurred to me back in school. But that's why Einstein rewrote the laws of physics and, thirty years after my school days, his question still leaves me scratching my head. However, as I look back and think about it, I am struck by the fact that none of my science teachers ever asked it either. Long and tedious, the science classes I attended didn't inspire much curiosity about the laws of nature or the workings of the universe. For me, like many red-blooded American boys, science classes were less than compelling—except for high-school Biology and the intense crush I had on my teacher.

There she was, talking about zygotes, chromosomes, and fertilization. But the problem was that I could not take my eyes off her as she sat on the edge of her desk, legs crossed, a white lab coat covering her dress. As if that distraction wasn't bad enough for a fourteen-year-old,

I was not "naturally selected" as a science student. In other words, while I managed to squeak through a year of freshman biology, it was a survival struggle of Darwinian proportions.

Sadly, as I look back at the sum of my scientific education before high school, I realize that my early science classes were equally forgettable. From Anatomy to Zoology, all that I retain of those endlessly dry and deadly dull classroom hours are vague recollections of having to draw bad diagrams of oversized eyeballs and eardrums and being forced to memorize the Periodic Table of Elements—an exercise that now strikes me as a futile and fairly useless piece of information. The only excitement came when the teacher's little experiments involving Bunsen burners went awry and produced an unexpected explosion and burst of smoke to the delight of the dozing students, eager for any sign of entertainment.

But even more remarkable than the high Boredom Quotient that I associate with science in my schooldays is the fact that I have no recollection of learning anything at all about space, astronomy, or the universe in my science classes. Apart from a single field trip to a traveling NASA display and our class trip to New York City to see *2001: A Space Odyssey* when it was all the rage—and which none of the teachers could quite explain to us—there is a large black hole where my education about the universe is supposed to be. We were all, as the popular television show of my childhood put it, lost in space. ("Danger, Will Robinson!")

What makes this severe shortcoming all the more noteworthy is the fact that my early education came during the most remarkable era in the history of space exploration. I was born in 1954, three years before the Soviet Union launched Sputnik—the basketball-sized satellite that changed American education and started the "Space Race." These were NASA's glory days—the era of *Life* magazine covers elevating John Glenn and the other astronauts to the celebrity pedestal now generally reserved for teenaged rock singers, talk-show hosts, and television sitcom stars. With Walter Cronkite as "cheerleader in chief," that national joyride in space climaxed with the Apollo Moon missions of the late sixties and early seventies, which captured the world's attention and imagination. I clearly recall the overwhelming public enthusiasm about space as our family stayed up late to watch lift-offs and Moon landings, proudly fulfilling President John F. Kennedy's challenge to put a man on the Moon. What an unbeliev-able rush of pride and near disbelief as we watched those grainy tele-vision images on July 20, 1969, from a quarter of a million miles away, as astronaut Neil Armstrong bounded down the last step of the lunar lander, *Eagle*, and told the waiting world, "That's one small step for [a] man, one giant leap for mankind."

But, inexplicably, that national excitement never translated into our classrooms. Bound by curriculum requirements and backbreaking but deadly dull textbooks, my teachers seemingly ignored the world outside our classrooms. In spite of the billions of dollars being spent on space exploration, the obvious fascination with outer space that made *Star Trek* the show we all watched, and basic human curiosity about the unknown, learning about the universe never made it into the lesson plans.

Even the obligatory field trips to the American Museum of Natural History in New York City managed to bypass the adjoining Hayden planetarium. And by the 1970s, when the planetarium got hip enough to run its Skyshows at midnight with Pink Floyd music play-ing, my attention was not exactly focused on sorting out the difference between the Big Dipper and the Little Dipper—which, I learned later to my dismay, were not even constellations but smaller star figures in the sky called *asterisms*. Obviously, like millions of other Americans, I didn't know much about space. The cosmos remained the exclusive preserve of the guys—and they were all guys back then—in white lab coats who knew exactly "what a slide rule is for."

This is a sad commentary on my schooling and education in gen-eral, because I am certain I am not alone. As with the other subjects I have covered in the Don't Know Much About series, the areas of space and astronomy hold considerable fascination for many people who have a basic curiosity about the universe. But, as a nation, we are "Astronomically Ignorant" in every sense of the expression. Textbooks, written by one set of professors and academics to be read by other pro-fessors and academics, left us miserable and muddled. Miseducation, media confusion, and Hollywood myth-making have all played signif-icant roles in creating this knowledge gap. And this celestial short-coming is all the more remarkable because the fascinating story of outer space and the universe is not just about physics and rockets and payloads. Basically, it is a human story.

After all, science started when people looked at the Sun, Moon, and stars and began to ask questions. Where did the Sun go each night and why did it return each morning? Why did the stars move so predictably across the sky? Why did the Moon move across the sky in such a regular way? What did the Moon and the tides have to do with each other? And what did *that* have to do with women and their cycles of fertility?

From these expressions of human curiosity came all the complexities of mathematics, astronomy, and the calendar, along with many of the myths and legends of human civilization. We now understand that many of the great structures of early civilizations—from the pyramids and ziggurats of the ancient Near East to Stonehenge, the Mayan temples, and imperial Chinese burial mounds—were designed with ancient astronomical calculations in mind. While there are still unanswered questions about many of these structures, those ancient calculations represented an early attempt to understand what was out there—and what comes after this life. In the human mind, those two concerns have always been connected. Since the earliest days of people's imagination, we have sought to make sense of our place in the dark, unreachable, and seemingly unknowable universe.

Unfortunately, that fascination with the heavens can mix with bad science with tragic consequences. In ancient times, the skies were blamed for all the good and evil that befell mankind. But a more glaring recent example of the danger of our imaginings about space came in March 1997, when members of a California cult called Heaven's Gate were convinced by their leader that they were aliens who would someday shed their earthly bodies. The tipping point for the cult came with the spectacular appearance of the Comet Hale-Bopp, which was visible for many months that year. Told by their leader the comet's tail concealed an alien mothership that was coming to take them away, a group of thirty-nine people committed mass suicide in a sad sacrifice to pseudoscience.

Still, this fascination with worlds beyond is clearly one reason that science fiction strikes such a nerve in the popular culture. Most of us experience science fiction with a combination of dread and delight. On one hand, we possess a deep-rooted, paralyzing terror typified by *The War of the Worlds* or the chilling paranoia of *Invasion of the Body Snatchers*. More recently, it has played out in the darkly threatening visions of *Alien* or *The X-Files*. On the other hand, millions of us have laughed over the lovable exploits of Steven Spielberg's *E.T.* and been dazzled by the majestic sense of awe in *Close Encounters of the Third Kind*. An entire generation has been absorbed by the childlike exuberance and timelessness of the *Star Wars* legend. We want to believe in our hearts, minds, and imaginations that something else is out there. From the ancients who charted the stars, to H. G. Wells and Flash Gordon, to *Apollo 13* and *Armageddon*, people have for centuries been transfixed by the heavens. We yearn to explore where no man has gone before, the "final frontier," as Captain Kirk called it on my childhood favorite, *Star Trek*.

It's a wonder, then, that most of us are so ignorant when it comes to the most basic facts about the heavens. Lots of people think that a white dwarf is one of Snow White's seven little friends. That nova goes great on bagels and cream cheese. That black holes have something to do with Calcutta. But this confusion doesn't mean that people don't care. It simply means that the education system has done a lousy job by turning something that is exciting and important into an exercise in tedium.

The good news is that it is never too late to fill in all of those black holes in our education. *Don't Know Much About the Universe* sets out to examine a subject about which there is tremendous ignorance and popular misconception, but even greater public fascination. It addresses these gaps in knowledge and still tries to appeal to that most basic human curiosity—looking skyward and wondering, "What's out there?"

A typical reaction among those who hear that I am writing this book is, "Wow! What a huge subject."

And there is no doubt about it. Writing about the universe is a task of, well, cosmic proportions. The numbers become boggling. Just considering the vastness of the universe hurts your hair. We live on one small planet orbiting one small star in one rather average galaxy.

But out there are some hundred *billion* galaxies, each containing *billions upon billions* of stars! How can you contain something that touches the infinite and put it all into one simple, accessible, and, ideally, entertaining book?

On the other hand, after writing a book about the Bible, taking on the universe seems a rather simple task. For much of the last two thousand years, people have been going to war over the Bible—a divinely inspired book about faith, tradition, and law that has been responsible for a good deal of the blood spilled in the Western world. With its competing interpretations and millions of passionate readers, the Bible has presented a minefield of controversy. So, taking on the vastness of space has a reassuring quality. Sure, there are scientists who argue about aspects of astronomy with the fervor others reserve for their religious faith. But on the other hand, no scientist has killed another over competing theories of the beginnings of the universe. Yet. (Although, I suppose, you could make the case that Giordano Bruno might be an exception. An Italian priest, Bruno was killed during the Inquisition in 1600 for his unorthodox views of the universe, among other heresies. More about him in part I.)

Unlike the Bible, with its competing schools of interpretation, translation, and authorized versions, when it comes to space, there is a body of facts—real scientific knowledge. And it grows larger by the day. We live in a time that is, in many respects, as revolutionary as the day some four hundred years ago, when Galileo trained a crude telescope on the Moon and stars and literally changed history. It is one of the most extraordinary periods in scientific history, as we have taken Galileo's primitive "spyglass" to other dimensions and given ourselves the capability to see far beyond the wildest imaginings of earlier generations of star watchers.

One of the chief reasons I chose this subject is the astonishing pace of discovery in the worlds of astronomy during the past few years. The Hubble Space Telescope. Deep-space probes. Advances in physics, such as the confirmation of the existence of tau neutrinos, the smallest particles of matter. All of these developments have either confirmed or altered long-held ideas about the universe. At a 1999 lecture I heard at the Hayden Planetarium, one speaker reported that more than half of everything ever written about astronomy has been produced since 1990—a reflection of the vast increase in information now available to researchers.

Much of that new information has altered basic ideas and accepted wisdom that has been taught for centuries. In other words, even if you were lucky enough to learn about astronomy twenty or thirty years ago, they've rewritten the books. One glaring example: Every school child learns that there are nine planets, right? Not so fast! They've demanded a recount in space. The very definition of a planet is being reconsidered, and some astronomers now cast doubt on whether Pluto—the ninth planet for most of us when we were growing up—is really not a planet at all! Did you grow up learning that there are three forms of matter—liquid, solid and gaseous? Then you haven't heard of *plasma*, an electrically charged form of gas that is actually a fourth type of matter and which makes up a large part of the Sun and other stars. (Just to avoid confusion, this plasma has nothing to do with biological plasma—the liquid portion of your blood.)

And you thought learning the "New Math" was a pain!

But there is another reason for this book. The world should be looking to the heavens with wide-open eyes as we enter the new millennium. (By the way, that's a completely arbitrary and very debatable dating concept that owes much to a monk who was trying to fix the calendar about six hundred years ago and who got some of his dates confused!) Not since those heady days in the late sixties, when John F. Kennedy's challenge to put a man on the Moon was taken up by the whole nation, has there been such promise and possibility in exploring space. Today, the powerful and ancient human preoccupation with discovery and pioneering in the skies has been reenergized by the wondrous robotic ramblings of *Sojourner* on Mars, the extraordinary images beamed back to Earth by the orbiting Hubble Space Telescope, and the initial steps taken in constructing the International Space Station. This football field–sized laboratory in space is being built and operated by a consortium of sixteen nations, and it may well become the real stepping stone to humanity's future in space exploration. We and our children will witness this incredible new phase in human curiosity and discovery—a prospect as exciting as being aboard one of Columbus's ships in 1492 or getting news reports from Lewis and Clark as they made their way west between 1804 and 1806, accurately gauging and reporting on the enormous size and spectacle of the American West acquired by Thomas Jefferson in the Louisiana Purchase of 1803.

Space and the cosmos are sure to become an even greater part of our lives, and the debate over how we explore it is an important public question that demands information. We have reached a day when a space station moves from the imaginings of science fiction writers like Arthur C. Clarke, to a reality we and our children will watch unfold. By early 2001, the United States and Russia were on the verge of what was once considered the unlikely vision of science fiction writers—allowing tourists to fly to the International Space Station. A wealthy investor and former NASA engineer, Dennis Tito, paid the Russian government $20 million to allow him to fly to the space station. Former astronaut Buzz Aldrin, the second man on the Moon, has formed a group to encourage space tourism in the next decade. There was even the suggestion that a *Survivor*-type reality television show might be set in an astronaut training center with the winner flying into space! The cash-strapped Russians, who planned further capitalist ventures, took their space age marketing a step further in May 2001. Pizza Hut reportedly paid the Russian government $1 million to have one of its pies delivered to the space station. Television pictures captured the pizza floating weightless as a happy cosmonaut gave a thumbs-up. Hold the pepperoni!

On a far more serious note, the nation and the world should be concerned as America debates the Strategic Defense Initiative, a planned antimissile system that involves billions of dollars being spent on a technically questionable venture that will affect our pocketbooks and international relations. In other words, thinking about space is no longer the exclusive turf of science fiction writers and whiz kids with pocket protectors who used to play in the backyard with rockets instead of baseballs.

To that end, *Don't Know Much About the Universe* is aimed at providing some basic information that will guide the sky searcher through the myths and confusions, providing an accessible road map to the "final frontier." White dwarfs, black holes, and dark matter. Novas and nebulae, quasars and pulsars. Asteroids, comets, and meteors. Solar systems and galaxies. The phases of the Moon. These are just a handful of the fascinating celestial mysteries that this book will cover as it seeks to answer old questions and inspire new ones. It sets out to "connect the dots" of what we know, providing the link from ancient times to the twenty-first century, showing how we have learned and what we can hope to discover.

In dealing with space and the vastness of the universe, I try to look at issues that go beyond the bounds of simple "Science 101" questions about the motion of the Sun, Moon, and stars. I like to ask the kinds of questions we may have wanted to ask back in school, but didn't have the nerve to. My lab-coated science teachers were certainly never big on creative thinking—or humor. So, in *Don't Know Much About* fashion, this book raises some questions of a sort not usually found in science textbooks:

- What does astronomy have to do with astrology?

- Did extraterrestrials build the pyramids?

- Who dug those canals on Mars?

- Is a "blue moon" really blue?

- Was Werner von Braun a war criminal?

- Is anybody else out there?

- Will we end with a bang or a whimper?

These kinds of questions are meant to spark the imagination and get readers thinking in new ways, examining the easy assumptions so many of us hold, or posing questions that many people never bother to consider. They include some seemingly irreverent or offbeat questions that open a back door into providing "serious" answers. And they are questions that move *Don't Know Much About the Universe* beyond astronomy into the broader fields of history, mythology, and, ultimately, cosmology. Along the way, I also hope to shatter any remaining myths and misconceptions you might still have about space. If you still think the Dippers are constellations, you are in for a not-so-rude awakening. If you think the Moon is made of green cheese, my work is really cut out for me!

Underlying the *Don't Know Much About* series is the idea that learning should be a lot more interesting than it was for most of us. "Education is not the filling of a pail but the lighting of a fire," wrote poet William Butler Yeats. And his words have become the unofficial motto of this series.

Learning works best when we toss out the old way of teaching— one subject at a time, all disconnected from each other—and bring different ideas and disciplines together. For instance, you can't understand the American Civil War by simply studying dates and battles, and ignoring geography, literature, economics, or religion. You can't understand the Bible without a grasp of archeology, ancient Near Eastern civilizations, anthropology, and literature. And you can't truly understand or appreciate astronomy and the physics of space without understanding mythology, religion, ancient history, and the history of ideas.

It is one thing to know that an astrolabe is a medieval device that was used to compute the location of stars. Another to discover that the same Geoffrey Chaucer who wrote *The Canterbury Tales* once wrote a how-to guide for constructing an astrolabe. It is a plain and simple fact that Galileo and Einstein both introduced ideas that changed our perceptions of the universe, as well as the course of history. But it is something else to discover that both of these pioneering thinkers had daughters born out of wedlock. Galileo's was sent to a convent and corresponded with her father. The fate of Einstein's daughter is something of a mystery. These aren't simply gossipy tidbits meant to sully their reputations; instead, I hope, it shows both of these men as human beings, which makes their accomplishments all the more remarkable.

Seeing how real people make a difference in history and understanding the connections between everyday life and historical events ultimately makes these subjects compelling and interesting. When you see Galileo as a proud, egotistical genius who occasionally took credit for things in questionable ways—a sort of Al Gore of the Renaissance—it not only makes him more human, but explains some of the difficulties he created for himself with powerful people in his day. Seeing the "warts-and-all" picture can give us a human handle on these historical giants, making it a lot more intriguing to learn about them and their achievements. Not only do I hope to give you some insight into Galileo, but many others—some familiar, some obscure—like Copernicus, Kepler, Tycho Brahe (a man with a metal nose), Newton, and others throughout the history of astronomy.

Don't Know Much About the Universe is organized into five main areas:

- **The Great Ocean of Truth** The book opens with a historical overview of mankind's fascination with space, documenting our progress in understanding the universe. This section will highlight a key theme throughout this book: how people in different times and cultures have viewed their place in the universe. It will also spotlight many of the geniuses of science who made extraordinary discoveries about space and the universe and how they did it, usually to laughter, ridicule, or worse from colleagues or the authorities.

- **Across the Gulf** Working from the Sun out, this section offers a guided tour of our solar system, including the Sun and each of the planets and their moons, along with all the other bits of celestial flotsam and jetsam like asteroids, comets, and meteors buzzing through the solar system.

- **Where No Man Has Gone Before** Moving beyond Earth's immediate neighbors, this section looks into the wider reaches of space: moving through our galaxy, the Milky Way, and beyond, into the vastness of the cosmos, focusing primarily on the other stars and galaxies in all their glory.

- **To Boldly Go** A brief overview of the extraordinary twentieth century race into space and the likelihood of life elsewhere in the universe.

- **The Old One's Secrets** The book's last section examines the larger questions that relate to the structure and nature of the universe, and, finally, the long-term fate of the universe.

Scattered throughout the book are the familiar features of the *Don't Know Much About* series: timelines that highlight the notable and obscure milestones in our understanding of space; and "Voices," a wide array of quotes from people throughout history, reflecting our changing views of the universe and our place in it. As in previous DKMA books, an extensive annotated bibliography offers additional reading.

Like all the other books in the Don't Know Much About series, this one is about questions. It is not meant to be the last word on the subject, but the "first word." I hope it will satisfy curiosity and provide answers, but also send readers out to ask more questions. Although it is meant to be a "down-to-Earth" book, it ultimately touches on subjects that can be, literally, way over our heads. It asks some questions to which there are no easy answers and sometimes no answers at all— yet!

It concludes at the uneasy intersection of science and faith, knowledge and belief. For centuries, these two opposing "religions" have battled each other for supremacy. But as we enter this next millennium, many people are seeking a way to bring them into some kind of harmony.

Don't Know Much About the Universe sets out to be an accessible road map through the mysteries of the universe, perhaps helping parents through the minefields of science homework along the way. Beyond that practical application, however, this book is also aimed at allowing us all to share in the age-old human wonder about the skies. When you are through, I hope you will be able to look up with understanding and appreciation—instead of a blank stare—and see that, in poet Walt Whitman's words, "every inch of space is a miracle."

WARNING: YOU HAVE PROBABLY READ AND HEARD IT BEFORE. BUT THIS IS SERIOUS! NEVER LOOK DIRECTLY AT THE SUN, EITHER WITH THE NAKED EYE OR ANY OPTICAL DEVICE. EVERY YEAR PEOPLE SUFFER EYE DAMAGE WATCHING ECLIPSES. THE ONLY SAFE WAY TO OBSERVE THE SUN IS TO PROJECT ITS IMAGE ONTO A PIECE OF WHITE PAPER USING BINOCULARS OR A SMALL TELESCOPE. LIGHT FILTERS HAVE TO BE USED TO SAFELY OBSERVE THE SUN.

PART I

THE GREAT OCEAN OF TRUTH

Who is it that darkens counsel by words without knowledge? . . . Where were you when I laid the foundation of the Earth? Tell me, if you have understanding. Who determined its measurements—surely you know! Or who stretched the line upon it? On what were its bases sunk, or who laid its cornerstone, when the morning stars sang together and all the heavenly beings shouted for joy?

The Book of Job 38:2–7

To me every hour of the light and dark is a miracle,
Every cubic inch of space is a miracle.

WALT WHITMAN, *Miracles*

What did we know about the universe and when did we know it?

What does astronomy mean and who invented it?

Who was the first astronomer?

Did Aristotle start the crystal craze?

Did anyone challenge Aristotle?

How did the Greeks get so smart?

Were the pyramids built by extraterrestrials?

Did the night sky look different in ancient times?

Is the Big Dipper a constellation?

Does cosmology have anything to do with a makeup kit?

What does an old pile of rocks in England have to do
with the universe?

What does astrology have to do with astronomy?

Why did St. Augustine give astrology such a bad name?

Did Hitler's astrologers get his horoscope wrong?

Was the biblical "Star of Bethlehem" an identifiable astronomical event like Halley's comet?

Who was Ptolemy and what did he have to do with "one thousand points of light"?

Why did Martin Luther call Copernicus a "fool"?

How did a sixteenth- century party animal who lost his nose in a duel change astronomy?

Who discovered how the planets move?

Who pierced Giordano Bruno's tongue?

Why was Galileo the "Al Gore" of the Renaissance, or, Who *really* invented the telescope?

Why did the Vatican arrest Galileo?

Did Newton's apple really fall?

INTELLIGENT BEINGS FROM OUTER SPACE

Remember Y2K? It seemed so important then, as we waited for the personal-computer Armageddon. Then came the election of 2000, the race that had us all wondering, "How come we can put a man on the Moon but can't count votes?" Perhaps, somewhere down the road, the remarkably close presidential contest of 2000 will loom large on the historical record. On the other hand, it could well turn out to be a historical footnote that will end up as a presidential trivia question on a future edition of *Jeopardy*.

So what did really matter in the year 2000? Maybe a child was born who might change the world as profoundly as Galileo, Newton, or even Hitler once did. Or, in a laboratory somewhere, an anonymous researcher had begun to unleash the secrets of Alzheimer's disease or the common cold.

Thinking along this same vein, take a look back at 1879, an otherwise unremarkable year. The European newspapers of the day probably gave substantial ink to the political machinations of Germany's chancellor Bismarck. Perhaps the odd experiments of a Russian scientist named Pavlov with a dog and a bell attracted some notice. And, certainly, the tinkering of a young man named Edison with some bits of filament and electrical wires must have drawn some interest among forward-thinking investors. But it is very unlikely that anyone besides his family took note of the birth on March 14, 1879, in Ulm, Germany of a boy named Albert. It isn't hard to argue that Einstein's birth was one of the most important things that happened that year. So much for the big moments in history as recorded by the media.

So what will they say about November 2000? What notable event will future generations look back on and say "This was where it all began"?

What if it turned out to be the first operations of the International Space Station, the orbiting laboratory, which opened for business even as presidential candidates George W. Bush and Al Gore were battling down to the wire? Perhaps for future generations, living and working routinely in orbit or in some far-flung space colony, this largely overlooked event may become the next millennium's equivalent of Columbus Day—a holiday that marks the beginning of a profound new era in human history.

If that scenario plays out, future historians may look back on the early years of the twenty-first century and ridicule the primitive equipment our astronauts on the International Space Station were using. They may marvel at how these "pioneers" were able to accomplish anything with such "crude" devices. Which is exactly how we look back at the remarkable discoveries of the past.

This opening section offers a look at the "history" of space—or, more accurately, reviews how humans have progressed in understanding the universe and our place in it. It shows how, over the centuries, people used reason and observation and eventually, a few crude inventions to figure out the universe and its workings. It shows how thinking about space and the universe went from superstition to myth to religion and, finally, to science. How people from the earliest days of civilization, staring at the seemingly uncountable stars, began to order the heavens. And how we moved, in a brief instant of cosmic time, from seeing the heavens as celestial pictures of bears, crabs, and archers, to comprehending the basic laws that govern the universe.

It is a remarkable story of human achievement and imagination. And much of this history is concerned with people who dared to question accepted wisdom—even when asking questions posed some danger. This is also the history of superstition and faith—which are different ways to describe things that science can't know for certain. We often take what we know for granted. But it was the work of many geniuses, sometimes working alone, sometimes working in remarkable groups, who have brought us to where we are at the beginning of the Third Millennium. And throughout this historical overview, the focus is on the "human face" of astronomy, especially in profiling some of the giants of science who profoundly changed the way we see the world.

However, it also points to some of the mistaken ideas and scientific blunders that have held sway, like some of those set forth by the great Aristotle himself—a great philosopher but a pretty lousy astronomer.

Those ideas shaped human thinking for centuries to come. Or those of Ptolemy, a Greek scholar of the first century whose correct ideas about the Earth and the universe were offset by his very wrong ideas about Earth's size and its being the center of the universe—with some amazing consequences for human history. This overview leads to the great thinkers of the Renaissance and Enlightenment, such as Copernicus, Kepler, Galileo, and, finally, Newton—men whose ideas and theories not only altered scientific understanding but changed the course of events in ways that few politicians or generals ever have.

VOICES OF THE UNIVERSE:
Greek philosopher ARISTOTLE (384–322 B.C.)

There is much change, I mean, in the stars which are over-head, and the stars seen are different, as one moves north-ward or southward. Indeed there are some stars seen in Egypt and in the neighborhood of Cyprus which are not seen in the northerly regions; and stars which in the north are never beyond the range of observation. . . . All of which goes to show not only that the earth is circular in shape, but also that it is a sphere of no great size: for otherwise the effect of so slight a change of place would not be so quickly apparent.

What did we know about the universe and when did we know it?

When you consider how old the universe is—a figure that ranges somewhere between 13 and 20 billion years—the human attempts to study the universe, and to try and gauge its age and size, are in a state of infancy. As a true science, astronomy started about 2,500 years ago—less than a finger snap in the life of the universe. And when you consider that telescopes have only been around for 500 of those years, astronomy is truly an infant. And when you consider that the first space flights came half a century ago, well, we're barely out of diapers when it comes to learning about the universe.

But long before people began to try and figure out the heavens using math and a few crude hand tools, humans have been studying the skies and fitting what they saw into some explanation for all the mysteries of life and the universe. For instance, thousands of years ago, in an ancient Japanese legend, the Sun goddess Amaterasu Omikami, the gentle source of all life, ruled the Plain of Heaven. The God of Storms was Susano, a spirit whose name meant Swift, Impetuous Male. Susano wanted Amaterasu, and his sexual advances on the Sun goddess were rude and violent. He broke her rice-field boundary and desecrated her temple with excrement. He killed and skinned a horse and threw its corpse into the hall where Amaterasu's handmaidens wove garments for the gods. In the modern parlance of the relationship guru John Gray, this was a classic "Men are from Mars, Women are from Venus" confrontation.

Angry at Susano's violent ravages, the Sun goddess withdrew to her cave in the sky. Her departure brought darkness, death, and chaos to the world. Demons spread doom and evil. The eight million spirits believed the universe would crumble without her golden light.

In an attempt to coax Amaterasu from her cave, a wise old god instructed another younger female goddess to perform a dance. As she danced, this nubile goddess began to loosen her kimono to the delight of all the assembled spirits. First she revealed her breasts, and then her belly. Finally her kimono fell open and she let it slip off. The shouts of laughter and delight from the entire company prompted the Sun goddess to peek outside and see what was happening. Once she appeared, the spirits quickly closed off Amaterasu's cave. Sunlight and order returned to the world.

This celestial striptease is a part of the ancient Japanese mythology that accounted for the changing seasons. As in the Greek myth of Persephone, the young fertility goddess who was taken to the Underworld each winter and allowed to return to her mother in the spring, Amaterasu's departure brought winter; with her reappearance, came the spring. Of course, the story of Amaterasu is a lot more intriguing than most seventh-grade science teachers' explanations of how the Earth's tilt on its axis causes the seasons to come and go.

Both the ancient Greek myth of Persephone and the Japanese legend of the disappearing Sun goddess, from whom the emperor of Japan is traditionally believed to descend, are typical of ancient efforts to explain the order of the universe and its mysterious connection to human life. In ancient Greece, the Milky Way was literally thought to be the breast milk of Hera, wife of Zeus, king of the gods. In order to attain immortality, the Greek hero Hercules, son of Zeus and a mortal woman, had to be given the breast milk of the mother goddess. The messenger god Hermes placed the infant at Hera's breast as she slept. But when Hera opened her eyes and saw the half-mortal child, she pushed the young Hercules away. It was too late. The milk that had begun to flow from her breast formed a trail in the sky—the Milky Way. Eventually, the lights of the Milky Way were thought by the Greeks to be the souls of the departed, an idea they may have borrowed from ancient Egypt, where the Greeks got quite a few of their ideas. Most ancient myths about the heavens and the gods attempted to set the universe into a human framework. Many of those legendary explanations were closely tied to life and death as well as fertility rites and human sexuality. But all of those mythical stories ultimately fell short.

The simplistic primitive explanations for why the Sun rose, or the Moon made its regular course in the night sky, or why the seasons changed or why the rains came and the Nile River flooded are shared by every human group. Rudyard Kipling called them *Just So Stories*—useful for explaining why the leopard got his spots, or why one star seems to stay in the same place in the northern sky, or why the Sun "rises" and "sets." But as people kept looking, learning, and thinking, human curiosity gradually pushed us from making up "Just So" stories to explain the universe to forming the earliest attempts to know the universe. As centuries went by and civilization progressed, superstition and religious faith were gradually replaced by reason and observation; logic and rational thought were used to try and order the cosmos. That is why astronomy is rightly called the First Science.

What does astronomy mean and who invented it?

Astronomy is derived from the Greek *astron* for "star" and *nomos* for "law," and it was first used by Greek thinkers and mathematicians about 2,500 years ago, during the remarkable burst of scientific, political, and cultural development known as Greeks' Golden Age, or the "Greek Miracle." Simply put, astronomy tries to explain the laws of matter in outer space, not just the stars, but all of the celestial bodies and phenomena. Yet, unlike the mythic stories that attempted to order the heavens, astronomy is a science. And science attempts to observe events, make reasoned explanations for those events ("theories"), and then test and retest those observations and explanations.

And while our modern minds find the ancient myths both fascinating and amusing, the fact remains that many of these primitive societies made extraordinary discoveries about the heavens. They began the long march away from superstition to bring us to telescopes, rocket ships, space stations circling the Earth, and manmade moons providing us with cell-phone service.

What the modern world likes to call "scientific thinking" originated in Greece, where an extraordinary collection of thinkers began to view the universe in a new way. Although the ancient stories of the Sun god Apollo driving his fiery chariot across the sky had been around for centuries, a group of Greeks began to put aside the myths of the past and explain the workings of the universe in terms of unchanging physical laws.

One of the first great bursts of scientific reasoning took place in Miletus, a city-state on the coast of what is now Turkey, in around 550 B.C., where a group of "Ionian philosophers" were among the first to believe that people could understand the universe using logic rather than mythology and religion. Driven by what was then a unique approach to the world, these mathematician philosophers began a search for a prime cause for all natural phenomena. What might seem like a simple idea for the modern mind to accept was remarkable at that time—they wanted to show that the personal forces of gods were not involved, only natural processes, in everything from the rising of the Sun to the disappearance of the Moon during an eclipse or the buildup of silt at the mouth of the Nile River.

One of the first of these great Greek thinkers was Thales of Miletus. Born about 625 B.C., Thales was an ancient Greek combination of Tom Edison and Bill Gates. A successful businessman with a mind for mathematics, Thales once supposedly cornered all the olive presses in the region and turned this olive-oil monopoly to huge advantage by renting the presses out at large profits. But, according to Aristotle, who came along about two hundred years after Thales, the Olive Oil King of Miletus was unconcerned with wealth. In *Politics*, Aristotle recounted, "He knew by his skill while it was yet winter that there would be a great harvest of olives in the coming year. . . . When the harvest time came, and many [olive oil presses] were wanted all at once, he let them out at any rate which he pleased and made a quantity of money. Thus he showed the world philosophers can easily be rich if they like, but that their ambition is of another sort."

In a move that might equate Bill Gates's leaving Microsoft to ponder the fate of the universe, Thales turned his attention from olive oil to practical science and is regarded as the founder of the Ionian school of natural philosophy. (What may be the only high-tech corporation named after an ancient Greek olive-oil salesman came into being when Thompson-CSF, an aerospace, defense, and information technology company, was renamed Thales late in the year 2000.)

One reason the Greeks began to do these remarkable things is that they became travelers—giving up the settled and very rocky prospect of simple farming and goat herding to become traders who sailed the Mediterranean, bringing them into contact with civilizations and ideas that had already been around for thousands of years. Having traveled to Egypt, Thales probably learned the craft of land surveying from the Egyptians, who had devised this form of mathematics to figure out who owned which property after each flooding of the Nile and who owed what to the pharaoh after each harvest. From this, Thales began to lay down some of the basic laws of geometry, later codified by Euclid. Legend has it that he calculated the height of the pyramids by comparing their shadows to the size of a stick. His "well-oiled" travels also took him to Mesopotamia, where he studied the skies, and it is thought that he predicted a solar eclipse, a feat that impressed the folks back home in Miletus so much that they decided not to go to war with a neighbor. Nobody is sure if this story is true or not, but it has lasted about 2,500 years just the same. Thales searched for a unifying principle or essence underlying all natural phenomena and he identified this essence as water. Thales believed that all matter came from water and that even the Earth had formed out of water, which was also the substance of stars. He imagined that Earth was a flat disk that floated on the cosmic ocean, surrounded by waters. The idea of a single material underlying all nature was the first "scientific" or rational attempt to explain the world without invoking the supernatural. Thales was essentially seeking the early Greek version of a "unified theory," one explanation for all of the universe, that physicists still seek today.

Then just who was the first astronomer?

The Greeks soon proved to be good at another idea—passing information along through academies and schools. One of Thales's pupils or followers was Anaximander (c. 610–545 B.C.), who introduced an early sundial and is believed to have written the earliest scientific book, now lost. Remarkably, he formulated a theory of the evolution of life, which said that life originated in the sea from the "moist element," which was evaporated by the Sun. To Anaximander, the presence of shells and marine fossils was proof that the sea once covered much of Earth and he theorized that humans must have originated in the sea and once resembled fish. But he also turned his gaze skyward and is often called the first astronomer. Anaximander believed that Earth was a cylinder—like a modern tin can—that floated freely in space, motionless, at the center of the universe. Mankind lived on one end of the cylinder. In his concept of the universe, the stars were fiery jets and the Sun a chariot wheel filled with fire, the highest of the heavenly bodies. He attempted to explain the heavenly bodies in the context of worlds constantly being created from the Infinite, only to perish and be reabsorbed into the Infinite, a very profound guess, as modern astrophysics has shown that the stars are essentially recycled stars (a subject to be explored in greater depth in part III). Though many of his ideas were mistaken, they were based on observation and rational argument instead of traditional superstition.

Following him was Anaximenes (585–526 B.C.), also of Miletus, who recognized that rainbows were a force of nature rather than a mystical event. He theorized that an infinite ocean of air, or vapor, was the most basic form of matter and supported the flat disk of Earth. For him, the Sun, Moon, and stars were masses of fire, all seen as disks that moved in cycles and attached to a rotating crystal sphere. Clearly, these early thinkers were mistaken in many of their notions, but they had begun the process of shifting the emphasis away from mythology and the rash actions of gods to a view of a natural universe ruled by laws.

This was a change taking place elsewhere in the Mediterranean world, including a Greek colony in what is now southern Italy, founded by a mathematician and mystic named Pythagoras (580–500 B.C.). Yes, you probably remember that name with dread because you were forced to memorize the Pythagorean Theorem, which says that the sum of the squares of the sides of a right triangle is equal to the square of the hypotenuse, or, "A-squared plus B-squared equals C-squared."

That is the sum-total recollection many people have of high-school mathematics. Pythagoras believed that the world and all of nature could be understood with numbers, and numerical and geo-metrical relationships acquired symbolic significance as part of the secret Pythagorean lore. Pythagoras is also credited with coining the term *cosmos*, to denote a well-ordered and harmonious universe—the opposite of the Greek concept of *chaos*. While they were grounded in the hard figures of mathematics, the Pythagoreans combined their reason with mysticism. Chief among their "mystical" notions was that the sphere—and a circle—was a perfect shape because every point on its surface was equidistant from the center. The long-standing influ-ence of this idea was the notion that the planets and other heavenly bodies moved in "perfect" circles around the Earth, which, Pythago-ras reasoned, was also a sphere. This mystic mathematician's recogni-tion that Earth is a sphere—some two thousand years before Colum-bus and Magellan sailed—was an argument that might have been lost except for the fact that the famed Aristotle later borrowed it.

Did Aristotle start the crystal craze?

Around 340 B.C., the great philosopher Aristotle took up the Pythagorean ideas and ran with them. He put forward some fairly convincing arguments for a round Earth in his book, *De Caelo* ("On the Heavens"). First, he noted that an eclipse of the Moon was caused by the shadow of the Earth when it was between the Sun and Moon. The Earth's shadow was always round, Aristotle reasoned. Second, Aristotle and the Greeks knew that the North Star appeared lower in the sky in the south than when seen in northern regions. That could only be if the Earth was curved rather than flat. Finally, he reasoned that Earth must be round because the sails of a ship come into view before the hull. If the Earth were flat, the entire ship would appear at once.

So much for Columbus and Magellan proving the world was round, just in case that bubble has not been burst for you!

That's what Aristotle got right. Some of his mistakes were more significant. Aristotle believed that the Earth sat motionless and that the Sun, the Moon, and the planets moved in perfect circular orbits around the Earth. This had little to do with his observations about the round Earth. Instead, it grew out of a nearly mystical belief that circles are the most perfect shape, an idea that Aristotle took from his mentor, Plato. In *De Caelo*, Aristotle stated that the Earth sits at the center of a great celestial sphere, made up of fifty-six successively smaller spheres. Each of these spheres carried a celestial body around the heavens in perfectly circular motion around the Earth. The sphere closest to the Earth contained the Moon.

The Greek concept of the universe, detailed by Aristotle, included the idea that something invisible had to hold up the stars and planets. Since the only hard, transparent substance then known was crystal, Aristotelian logic insisted that spheres of crystal hold the stars, the Sun, the Moon, and the planets in place. Aristotle called this crystalline substance *quintessence*, "the fifth element," along with earth, fire, air, and water. The Greek tradition also included the idea that the heavens are very different from Earth. On Earth, everything is constantly changing, in a state of transformation. The heavens were eternal and incorruptible, where nothing ever changed—except for comets. Since comets appeared in the heavens, lasted a few days or weeks, and then disappeared, that constituted change. Comets, Aristotle argued, must be earthly rather than heavenly, and must be some sort of spontaneous fire in the upper atmosphere caused by "exhalations" from Earth. (In fact, comets are essentially dirty "snowballs"—

collections of frozen dust, whizzing through space, trailing dust and gas, which burn up but leave those remarkable tails we can some-times see on Earth. More about them in part III.) All of these heav-enly motions, to Aristotle, were produced by the Prime Mover, who acted outside the starry vault, and who, in later Christian times, was identified as God.

Although Aristotle died in 322 B.C. at the age of sixty-two, his stu-dent Alexander the Great took Aristotle's ideas wherever his armies went. Greek knowledge, ideas, culture, language spread through Egypt and Palestine across Asia Minor to present-day Afghanistan and Pakistan, nearly to India. And although Alexander died within a year of Aristotle, the old philosopher's concepts, planted with the banner of Alexander the Great, would hold sway in most of the civilized world for centuries to come.

Did anyone challenge Aristotle?

Aristotle's thinking dominated philosophy and science for nearly 1,900 years, and his ideas about physics remained unchallenged until the Middle Ages, when scholars began to question and refine his assumptions. The impact of this Greek philosopher was so powerful that even the early Christian Church was influenced by his thought. But while Aristotle's ideas were to become "gospel," there was at least one voice opposing his. Aristarchus of Samos (310–230 B.C.) was the first person known to suggest that the Earth might orbit a stationary Sun. Little else is known of Aristarchus and, needless to say, his ideas were dismissed. The idea of a moving Earth seemed implausible, and people who had no knowledge of gravity—the force that keeps our feet on the ground—scoffed for centuries that if Aristarchus was right, what kept things from being spun off the Earth and flying into space?

VOICES OF THE UNIVERSE:
Greek scientist ARCHIMEDES (c. 287–212 B.C.)

Aristarchus of Samos brought out a book consisting of some hypotheses, in which the premises led to the result that the Universe is many times greater than now so called. His

> hypotheses are that the fixed stars and the Sun remain
> unmoved, the Earth revolves about the Sun in the circum-
> ference of a circle, and the Sun lies in the middle of the
> orbit.

Although his contribution is more fittingly tied to Earth, the efforts of another Greek (or Libyan who lived in Egypt) deserves mention as they demonstrate how these remarkable Greek thinkers were applying reason and geometry to understanding the world. Eratosthenes was the librarian at the famed collection in Alexandria, who once measured the Earth using little more than a stick and some geometry. Knowing that the Sun was almost directly overhead in Syrene (modern Aswan), Egypt, on the summer solstice, and so cast no shadow at noon, Eratosthenes measured a shadow cast by the Sun on the same day in Alexandria. Comparing the distances between the two places, he used the length of that shadow to calculate that the distance from Syrene to Alexandria must be 0.02 of the circumference of Earth. Using ancient measurements that have been adjusted to modern standards, Eratosthenes was able to calculate the circumference of the Earth to an astonishing degree of accuracy, a result that has been translated in modern measurements as 24,608 miles (39,690 kilometers), remarkably close to the modern value of 24,901.55 miles (40,075.16 kilometers). Having calculated the circumference of Earth, Eratosthenes then also figured the diameter of Earth as 7,850 miles (12,631 kilometers), again remarkably close to today's accepted value of 7,926.41 miles (12,756.32 kilometers).

Finally, in the pantheon of Greek skywatchers comes Hipparchus (c. 146–127 B.C.). He was born in Nicaea, in the northern part of Turkey. Almost nothing is known of his life, but he is considered the father of "systematic astronomy" and perhaps the greatest astronomer of the ancient world. Depending on your high-school career, you might want to curse him for also inventing trigonometry. In his spare time, Hipparchus devised the astrolabe, a device for measuring the positions of objects in the sky, which was the chief tool for astronomers until the telescope came along some 1,600 years later. He combined the vast astronomical information borrowed from Baby-

Ionia, which was then a conquered nation under the rule of Alexander's successors, combining them with his own for the most comprehensive table of star charts ever compiled. It contained about 850 entries, and designated each star's celestial coordinates, indicating its position in the sky. Hipparchus divided the stars according to their brightness, or magnitude, a standard still employed today.

All this Greek brilliance—okay, some of it was from Libya and modern Turkey—didn't happen in a vacuum. The Greeks were amazing scientists, politicians, sailors, merchants, playwrights, and sculptors, to mention a few of the other highlights of their golden civilization. But they had help. All of their wanderings around the Mediterranean world had introduced them to the ideas and achievements of other ancient civilizations—particularly those of ancient Mesopotamia and Egypt, where the pyramids had already stood for centuries before the Greeks came along.

How did the Greeks get so smart?

Most of us learned about the heights of Greek civilization back in school. They told you it was the birthplace of western civilization about 2,500 years ago. All of this Greek science that flourished for a period from about 600 B.C. to the time of the first century A.D.—the beauty of their achievements in math and science—are only a fraction of an era that has been called the "Greek miracle." Centuries ago, the Greeks reached a pinnacle of achievement in government, science, philosophy and sculpture, poetry and drama. They invented the writing of history, and western theater. But what the school books never really explained was how a small group of goatherders and olive farmers built one of the most extraordinary cultures in human history.

They had less wealth and land than the Egyptians or Babylonians. They lacked large numbers. Then what prompted this revolution, particularly in science? Some historians and anthropologists contend that virtually every culture, if left to its own devices, would eventually "discover science. This is the old argument that says, if you give enough monkeys enough typewriters, they will eventually produce the complete works of William Shakespeare. Maybe, but we wouldn't want to wait for that. Certainly with a bigger head start, the Babylonians and Egyptians did not accomplish what the Greeks did.

One explanation is that the very wealth and relative ease of life for the Egyptians held them back. They had no incentive to innovate. It's the historical and cultural equivalent of "no pain, no gain." When you don't have to work hard to make a living, when things come too easily, you get lazy. Life on rocky, small Greece, a loose collection of minor cities struggling to eke out an existence, was far tougher. As Carl Sagan wrote in *Cosmos*, "Some of the brilliant Ionian thinkers were the sons of sailors and farmers and weavers. They were accustomed to poking and fixing, unlike priests and scribes of other nations, who, raised in luxury, were reluctant to dirty their hands. They rejected superstition, and they worked wonders." This is the necessity-is-the-mother-of-invention approach to history and it is a valid one.

Another key was the ocean. With poor land and sparse rainfall, the Greeks were forced to turn to the Mediterranean. Unlike Egypt, which was tied to the Nile, or Babylon, where life was circumscribed by the Tigris and Euphrates Rivers, the Greeks were forced to reach outward for survival. While trade was important to both of those earlier ancient civilizations, for the Greeks it was a form of economic survival that became a way of life. Trade is not only profitable, but it increases the exchange of ideas. The Greeks built a Mediterranean trading empire that was one key to their commercial success. Once you don't have to struggle to eat, there is time to think. The Greeks became a great trading and military sea power, for the same reason England and Japan would later become empires—all three small island nations with limited natural resources, forced to reach out to the world through trade. For the Greeks, those contacts enriched the traders and provided new ideas.

Of course, Greece was not without flaws or problems. Women were second-class citizens. And Greece eventually became a slave culture, which, many historians agree, helped to bring about its downfall. Once a culture grows dependent upon a slave class, the incentive for invention and discovery is lost.

As I looked, a stormy wind came out of the north: a great
cloud with brightness around it and fire flashing forth con-
tinually, and in the middle of the fire, something like gleaming amber. In the middle of it was something like four living
creatures. . . . Each moved straight ahead; wherever their
spirit would go, they went, without turning as they went. In
the middle of the living creatures there was something that
looked like burning coals of fire, like torches moving to and
fro among the living creatures; the fire was bright, and lightning issued from the fire. The living creatures darted to and
fro, like a flash of lightning.

As I looked at the living creatures, I saw a wheel on the
Earth beside the creatures, one for each of the four of them.

Were the pyramids built by extraterrestrials?

In a series of immensely popular books, author Erich von Däniken
has put forth the notion that Earth was visited thousands of years ago
by extraterrestrial creatures in spaceships. His most famous book,
Chariots of the Gods (1968), which has sold millions of copies worldwide and been published in nearly thirty languages, relies largely
upon similarities among certain legends, primitive artwork found at
diverse locations, and ancient structures that belong to different civilizations that had no means of communicating with each other. A
carving on a Mayan tomb lid is said to show an astronaut at the cockpit of a spacecraft. An aboriginal cave painting in Australia supposedly
depicts a figure wearing what could be a space helmet. Ancient diagrams in the Peruvian plain of Nazca might be an alien landing strip.
In von Däniken's view, these aliens were not the malevolent invaders
of countless Hollywood science fiction thrillers, but great builders.

These alien builders, to their remarkable credit, were responsible
for the great stone megaliths at Stonehenge in England, temples of
Central and South America, Mesopotamian ziggurats, the colossal

statues found on Easter Island in the Pacific, and the Egyptian pyramids, among other ancient mysteries. While von Däniken argues that
these aliens were primarily builders of the very first order, they could
also be destructive. One act of destruction von Däniken attributes to
these ancient alien visitors is the cataclysm that reduced Sodom and
Gomorrah to ashes in the story of Lot found in the biblical Book of
Genesis. Many people, von Däniken among them, also point to the
biblical vision of the Hebrew prophet Ezekiel (see above) as further
evidence of ancient "close encounters," a notion that continues to
inspire Hollywood visions from episodes of *The X-Files* to science-
fiction films like *Stargate*.

From the vantage point of pure science, there is, of course, no
"proof" of any such alien visitation. The evidence that believers point
to mostly consists of mysterious objects, drawings, and sculptures from
ancient times that have generated heavy doses of conjecture. Lacking
sufficient information or clearly supportable facts, many people prefer
to turn to alien giants to explain away ancient mysteries.

What is clear from history, however, is that there was a burst of
extraordinary development around the globe at approximately 3000
B.C. For instance, at this time in ancient Mesopotamia—the famed
"cradle of civilization" to nearly every schoolchild—people developed
agriculture, writing, the potter's wheel, and sailing ships. They also
invented beer. These were the people who lived on the plain between
the Tigris and Euphrates Rivers—the word "Mesopotamia" is from
the Greek for "between two rivers." By 2000 B.C., this fertile land
between the rivers in what is modern Iraq was home to a sophisticated
civilization that had begun to look to the heavens for guidance.
Ancient Sumerian writings attributed their knowledge to "gifts of the
gods," and some people have interpreted that to mean alien visitation.
You might choose to accept that theory or place it on the pseudoscience spectrum somewhere alongside astral traveling and past-life
regressions. But what cannot be avoided is the fact that this persistent
fascination, or obsession, with alien encounters from ages past speaks
to something of the human mind—an intense human desire to know
what we cannot really see. And that was especially true when it came
to the heavens.

Did the night sky look different in ancient times?

In the modern world, in which most people live clustered around brightly lit metropolitan areas, in which smoke and other industrial pollutants have combined to screen out most of the visible starlight, the wonder of the night sky has largely been lost. If you can find an unpopulated and less brightly lit section of the country, preferably at higher altitudes, you begin to sense the extraordinary spectacle each night's sky must have presented to the people of ancient times. Their view of the sky was unaffected by artificial lights and pollution. With no television or video games—or even books to read—people had a little more free time to contemplate the skies. The regular display of shooting stars and moving canopy of stars that most of us never get a chance to observe, would have been bright and beautiful—often frightening. Imagine how incredible it must have been!

What they saw over the course of centuries was that the sky above them was a dome of lights—often referred to as the *celestial sphere*—whose movements were regular and predictable. The Sun's routine was simple to figure out. Each day it rose in the eastern sky and set in the western sky. So, too, was the Moon's regular pattern. But even the seemingly countless lights of the night sky had fixed paths. This pre-occupation with the stars and planets soon began to be reflected in almost everything that the ancients built or conceived, from their religious beliefs to the idea of a calendar to their great architectural wonders.

The Mesopotamians are the earliest civilization known to have actively studied the stars and planets, and produced the first known charts of the heavens. Like other ancient civilizations, the Mesopotamians looked to nature for signs of things to come, whether it was reading the entrails of sheep or looking at the heavens. When they looked to the stars, the Mesopotamians were first concerned with the seasons. Was it possible to tell from the night sky when the rains would come again? The Sumerians, the earliest of the Mesopotamian cultures, were the first to record the movements of the skies and they did so for more than seven centuries. With that long collection of information to guide them, they placed the cyclical patterns of the

Moon, then the Sun and the planets into an orderly cyclical rhythm. In other words, they invented the beginning of the earliest calendar. Using the calendar they inherited from the Sumerians, the Babylonians later devised seasons, months, and, using a very advanced number system based on sixty, divided the day into twenty-four hours, the hour into sixty minutes, and the minute into sixty seconds. The Babylonians also gave the week seven days, which they named after the Sun, the Moon, and five bright "stars" that were actually planets. The vestiges of this Babylonian concept survive in modern times in the names of the seven days of the week. In other words, these people of the ancient Near East invented beer, then they invented the weekend.

The daily observation of the Sun and Moon became the key to predicting patterns—especially when agriculture began about five thousand years ago in ancient Mesopotamia and Egypt. Planting crops and the religious festivals that eventually grew up around them were completely tied to the passage of the Sun and Moon—the beginning of the calendar that marked human civilization. One of the easiest ways to mark the passage of time was the twenty-nine- or thirty-day cycle between full moons. A "year" equaled the twelve successive "months." The English words "moon" and "month" are derived from the Latin *menses*, for month, as is the word "menstrual." In an attempt to connect the natural world with life, some of the ancients made a mystical connection between the Moon's regular passage and the monthly fertility cycles of women. Over time, the Moon came to be viewed as a symbol of fertility—the key to life in a world of superstition.

But just as it has very little to do with human biology, the moon is an imperfect timekeeper. A lunar calendar based on the phases of the Moon is not a very accurate way of figuring a "solar year"—the time in which Earth completes one orbit of the Sun, or approximately 365.25 days. A lunar year of twelve months adds up to only 354 days. Many lunar calendars were adjusted by the addition of a month, as is still true of the Jewish, Islamic, and Chinese calendars. In ancient Rome, the priests controlled the calendar, and extra months could be added quite arbitrarily—often by payment of a bribe. But as agriculture grew more important to these early farmers, a more accurate way of calculating the arrival of planting and harvesting times became a practical necessity.

Gradually, people realized that the seasons were more closely tied to the movement of the Sun and stars than that of the Moon. And the first place where a calendar based on the solar year was created was ancient Egypt, home of an elaborate solar cult. This bears no resemblance to modern sun worship, which in St. Tropez and Southern California consists of slathering up with oil and hitting the beaches for a few hours. The Egyptians created an entire society and religion around Sun worship, featuring many gods who represented the Sun at different times of the day. Four thousand years ago, the Egyptians believed that a sacred boat carried the Sun god Ra (or Re, as he is also known) across the sky each day. To explain why Ra's boat rose over the eastern horizon and sank below the western horizon at night, a complex mythology emerged. The Egyptians believed that Ra's nocturnal journey was a passage through the underworld, confronting demons and dangers. Each night, this most powerful of gods was able to defeat the forces of darkness and return triumphant the next day.

It is easy to understand the centrality of the Sun in a remarkable civilization that sprouted on the thin strand of farmable land alongside the Nile River, the world's longest. The Egyptians were completely reliant upon the annual flooding of the great river. When the Nile's waters rose and inundated the banks of the Nile and surrounding territory, the cropland was watered, and the rich soil that was left nourished Egypt's wheat and other crops. Obviously, this regular flooding, tied as it was to the food supply, became momentous to these people. Eventually, ancient Egyptian skywatchers realized that the star Sothis—which the Greeks and modern astronomers call Sirius (the dog star)—appeared on the horizon at dawn just before the Nile reached flood stage each year. By measuring the time from one rising of Sothis (Sirius) to the next, the Egyptians produced a calendar with a year of 365.25 days—the modern year.

The daily track of Ra and the importance of the flooding of the Nile made the Egyptians careful observers of the heavens. The pharaohs who came to lead Egypt were not dumb. As the Egyptian civilization grew more sophisticated, they cultivated the notion that the pharaoh was also a god, descended directly from Ra. In Egyptian belief, the Milky Way was a heavenly counterpart to the Nile, a river of stars along which the departed pharaoh took his place among the other gods of the sky. In time, the pharaohs would need great tombs, and the concept of the pyramid evolved not simply as a resting place and memorial for the pharaoh, but as a stepping-off point from which he could begin his journey to the heavens.

The ruins of thirty-five major pyramids still stand near the Nile River, each built for the body of an Egyptian monarch. The pyramids clearly show an astonishing sophistication in their design and construction. Their grandeur, which has been partially eroded by centuries of wind and weathering and, more recently, pollution, and the technical demands of raising these monuments with primitive tools and simple machinery, is still boggling to modern students of the pyramids. The bases are aligned with geometric accuracy that defies the idea that they were a product of an ancient civilization with no modern measuring devices. They are also completely connected to the Egyptian view of the universe, a view closely tied to their astronomy. The four sloping sides of a pyramid, for instance, suggest the slanting rays of the Sun, by which the king's soul could rise to join the other gods in the sky.

In recent years, many pseudoscientific and supernatural beliefs have been linked to the pyramids, including *Chariots of the Gods*, a theory of extraterrestrial intervention. Erich von Däniken suggested incorrectly that the Egyptians did not have enough trees, rope, or manpower to build the pyramids. All three of these ideas have been thoroughly refuted. Egypt had access to plenty of trees, and Egyptian reliefs even depict builders using rope. As to manpower, the pyramids are now known to have been built with the conscripted labor of farmers during the Nile's flood stage. Egyptian people, not slaves, willingly worked a part of each year to construct the pyramids and other Egyptian landmarks. (The pyramids were also built long before the earliest possible dates for the arrival of the biblical children of Israel, an event which is not recorded in any Egyptian records.)

Another recent theory that captured public attention was the notion that the pyramids were laid out to form a map of the stars, specifically the stars in an asterism known as "Orion's belt," in the constellation known as Orion, the Hunter. The Egyptians connected this grouping with Osiris, the god of resurrection. This theory suggests that the three great pyramids at Giza were all built to align with these three stars, one of which was a star the Egyptians called Thuban. However, the last time these stars in Orion matched the alignment of the pyramids was more than twelve-thousand years ago. In November 2000, a researcher using the orientations of stars that aligned with the pyramid's base, was able to date the start of the construction of the Great Pyramid at Giza to 2478 B.C., a date well within the accepted notion of when the pyramids were built—by Egyptians, not super aliens.

Is the Big Dipper a constellation?

Modern astronomy recognizes eighty-eight areas, called *constellations*, into which the sky is now divided for the purposes of identifying and naming celestial objects. Some of these are familiar as the twelve signs of the zodiac still used in popular astrology. These ancient constellations were groupings of stars identified by the Babylonians, the first to connect celestial patterns of stars with figures in their mythology. They also observed that these stars moved in a regular pattern, along with the Sun and Moon, along a path in the sky that is now known as the *ecliptic*. The Greeks adopted this concept from the Babylonians and then added to it, to include Greek mythological figures such as the great hero Hercules; Andromeda, who was chained to a rock and left to be devoured by a sea monster; and Pegasus, the winged horse.

Today, the concept of a constellation is a convenience, a direction in which to point when looking at the sky. Although the stars seem to be close and somehow related, we cannot tell simply by looking that they may be separated by hundreds or thousands of light-years, the standard measurement of distance between stars. Over the centuries, many more constellations were recognized, and the sky chart was expanded, particularly after the sixteenth- and seventeenth-century voyages of discovery opened up the Southern Hemisphere where the view of the sky is very different from the Northern Hemisphere.

Within these large constellations are often found smaller groupings of stars, often in easily identified shapes called *asterisms*. The Big Dipper, for instance, perhaps the most recognizable grouping of stars, is an asterism within the constellation Ursa Major, the Great Bear. The

Little Dipper is another asterism, found within the constellation Ursa Minor, or the Little Bear.

Does cosmology have anything to do with a makeup kit?

There is another name for all of this looking up and studying the skies. While astronomy concentrates on the movements of the heavenly bodies, *cosmology*, again from the Greek, and meaning "to order the universe," attempts to place the movements of the stars and planets into some kind of order. Four thousand years ago, the Babylonians were able to predict the apparent motions of the Moon, stars, planets, and the Sun, and could eventually predict eclipses. In discussions of "cosmology" with respect to ancient civilizations, the word has a broad meaning that relates to how these civilizations understood the world to work—it went beyond the actual movements of the stars and into their entire religious beliefs and understanding of creation and life itself.

When the Greeks later borrowed heavily from the Babylonian knowledge of the heavens, they attempted to construct their own "cosmological model," a means to explain the motions of the heavens. When they introduced science into the equation, they moved cosmology away from superstition and belief. Just to straighten things out, cosmetology, the art and science of using beauty products, does come from the Greek *cosmos*, for "order." In essence, cosmetology is a way to arrange, or bring "order" to, the face.

In the fourth century B.C., the Greeks developed the idea that the stars were fixed on a celestial sphere, which rotated about the spherical Earth every twenty-four hours, and the planets, the Sun, and the Moon moved in the "ether" between the Earth and the stars.

All of this skywatching was not confined to the ancient Near East. Every other ancient civilization looked heavenward, attempting to use the skies to order their worlds and creating mythologies around those observations. What the people of the ancient Near East and Mediterranean couldn't know was that the Chinese were not far behind in their studies of the heavens. Operating thousands of miles to the east, the Chinese were beginning to examine and chart the skies at about the same time as the Babylonians. Systematic observations of the heavens had begun around 3000 B.C. in China, where they had also computed a 365-day year and were extremely sophisticated in their ability to predict eclipses. The Chinese had identified constellations, sometimes in ways similar to the Babylonians. For instance, the Chinese and Babylonians both identified the constellation still known as Draco as a dragon.

What does an old pile of rocks in England have to do with the universe?

Another remarkable landmark of ancient cosmology—or, for the alien-encounter camp, evidence of visitors from space—was contributed by the stargazers of ancient Britain. About three thousand years ago, using picks made from antlers, they began the construction of what is known as Stonehenge, another site that has produced widespread speculation over the centuries. (Its name comes from Old English "stanenge," literally, "stone hinge.") Built and rebuilt in stages for two thousand years—until about 1100 B.C.—it is a complex assembly of boulders, ditches, and great upright stones, located in southwest England. Stonehenge has inspired tremendous speculation as well as serious scientific investigation. The public fascination with this place of ancient mystery grew in recent years to the point that it had to eventually be cordoned off with barbed wire and electric fencing by the British government, to prevent its destruction by vandals as well as the fascinated thousands who have visited the site either out of curiosity or more supernatural interest. Over the past three thousand years, erosion, souvenir hunters, and vandals have altered Stonehenge. What remains today is only a partial outline of its original structure. Recent studies indicate that it probably once had thirty or more tall sandstone blocks, arranged in a circle. Lying horizontally on top of these were smaller stones, making an unbroken circle. A second inner circle of stones enclosed a horseshoe-shaped group of still smaller stones. Today, only a partial outer ring and a few inner stones remain.

The monument's builders are widely thought to be druids, an ancient priestly class, and its function, while still mysterious, was clearly related to astronomy. It is intriguing to imagine that this ring of stones, whose very existence is difficult to explain, is some sort of "ancient observatory." Parts of Stonehenge are aligned with the positions of the Sun and Moon on the *solstices* (Latin for "still sun"). The solstices are the days that mark the longest and shortest days of the year—obviously important days to Sun worshipers as well as early farmers who had no calendars but still had to devise some way to figure out a good time to plant their crops. By standing at the center of the ring of stones and looking northeast, one could observe the first rays of the Sun on the summer solstice lining up with one stone called the Heel Stone.

The exact purpose of Stonehenge and even the intricacies of its construction, while mysterious, still leave no suggestion of alien builders. But its fascinating allure to modern generations is a stark reminder of the connections people still feel to the cosmos. When we speak of druids or ancient astronomers, the word *cosmology* is used loosely to mean their overall view of the heavens. However, in the modern sense, *cosmology* means the scientific study of matter in outer space, especially the positions, dimensions, distribution, motion, composition, energy, and evolution of celestial bodies and phenomena. It is distinctly a mathematical and theoretical world far removed from that of druids and Egyptian priests. Yet their ancient fascination remains for us today. But not always with the "pure" light of science. Belief, superstition, faith, and curiosity still haunt our dreams of the skies.

VOICES OF THE UNIVERSE:
WILLIAM SHAKESPEARE

This is the excellent foppery of the world, that, when we are sick in fortune—often the surfeit of our own behavior—we make guilty of our disasters the Sun, the Moon, and the stars; as if we were villains by necessity, fools by heavenly compulsion, knaves, thieves, and treachers by spherical predominance, drunkards, liars, and adulterers by an enforced obedience of planetary influence. (*King Lear*)

The fault, dear Brutus, is not in our stars, But in ourselves . . . (*Julius Caesar*)

What does astrology have to do with astronomy?

A relative handful of American newspapers carry a stargazer's chart every day. With it, you can go outside on a clear night and watch the heavens as they wheel around the Earth. But for most people, this astronomical helper is as foreign, or arcane, as the pork-belly and soybean futures tables published in the business section.

At the same time, far more newspapers and magazines carry another type of star chart every day, one that is carefully followed by millions of readers—the daily horoscope—found under names such as "Love Signs" or "Sun Signs," which have made famed astrologers, including Linda Goodman and Jean Dixon very wealthy people. Most newspaper editors will tell you it is in their papers for entertainment purposes only, and many people give the horoscope a quick glance for the fun of it, the way they read the fortunes in their Chinese cookies. On the other hand, millions of people take these daily "zodiacs" very seriously, turning to them for insight into their decisions and their futures.

The belief that the positions of the Sun, Moon, and planets influence the course of human affairs and earthly occurrences is as old as the fascination with the skies. From the earliest days of studying the skies, astrology and astronomy were closely linked. As a "science," astrology was utilized to predict or affect the destinies of individuals, groups, and even nations. The Greeks built upon existing Mesopotamian and Egyptian lore while applying the new rules of geometry to describe the orbits of the planets. They also fully developed the idea of the twelve constellations of the zodiac and the concept that the positions of the planets, Sun, and Moon at the time of birth determined a person's fate, which gradually became an accepted idea. It grew to include the notion that the heavens also governed wealth, marriage, and death. Even Greek medicine, which made tremendous scientific strides, saw various functions of the body and different organs as being ruled by different combinations of planets and constellations.

The Romans, who later adopted or adapted many Greek ideas, preferred their own ancient means of divining the future, but Greek astrology eventually became accepted in Roman times as well.

Emperor Augustus elevated astrology to a royal art. And the Roman writer Seneca, who was born around the time of Jesus and later tutored the notorious Emperor Nero, once said, "Think you so many thousand stars shine on in vain? . . . Even those stars that are motionless, or because of their speed keep equal pace with the rest of the universe and seem not to move, are not without rule or dominion over us."

Why did St. Augustine give astrology such a bad name?

In the Christian era, astrology started to take some lumps. The most influential church thinker of the Middle Ages, St. Augustine (354–430), summed up the Christian objections to astrology. In his *Confessions*, Augustine recalled that before he renounced his pagan upbringing, he had consulted astrologers. But he was convinced otherwise and, to him, "the lying divinations and impious dotages" of astrologers were the Devil's work. *In City of God*, he said that the world is ruled by divine providence, not by chance or fate. The proof for Augustine came in a story told by a friend. Two children were born at the same moment, in the same household. Surely their fates would be the same. But one of the infants was the child of the master of the house. The other was the child of a slave. The position of the stars, exactly the same at the moment of their births, did not change their fates. The same logic would apply to the birth of twins, and Augustine pointed to the biblical brothers Esau and Jacob, born instants apart yet so completely different in temperament and fate.

In spite of Augustine's objections, astrology continued to flourish through Christian times, and many popes over the centuries called upon astrologers in making decisions. Even the arrival of the Renaissance and Enlightenment did not weaken its pull. In fact, throughout the early history of astronomy, many of the greatest names in the field, responsible for some of its greatest discoveries, turned to astrology to pay the bills. Copernicus, Tycho Brahe, Kepler, and Galileo provided clients with astrological charts. Even the mighty Newton is said to have first become interested in astronomy after looking at an astrological book. (All of these astronomers are discussed later in this chapter.)

They had been preceded by great scholars of the Middle Ages, such as Albertus Magnus and St. Thomas Aquinas, the influential church thinker, who both admitted that the stars had a strong governing influence. But they argued that man's free will allowed him to resist these powers.

As a "pseudoscience," astrology must be viewed as diametrically opposed to the findings and theories of modern Western science. As the medieval Jewish philosopher Maimonides (1135–1204) once put it even more strongly, "Astrology is a disease, not a science."

Did Hitler's astrologers get his horoscope wrong?

In spite of science, astrology still holds a particular fascination for many people who see their lives linked to the constant motion of the stars. The late Carl Sagan pointed out in his popular book *Cosmos* that modern language retains a sense of the influence of astrology. As Sagan noted, "*Disaster*, which is Greek for 'bad star,' *influenza*, Italian for [astral] 'influence,' *mazeltov*, Hebrew—and, ultimately, Babylonian—for 'good constellation.'"

Well documented is Hitler's preoccupation with astrology. It has been said that on the basis of what his astrologers saw in his star charts, Hitler squandered several opportunities to attack England at a time when such an invasion might have succeeded. In fact, during World War II, the British government used astrologers to attempt to figure out what Hitler's astrologers were telling him!

The basic principle of astrology is that the heavenly bodies somehow influence what happens on Earth. Astrologers demonstrate this influence by means of a chart called a *horoscope* (from ancient Greek for "hour" and "observer"), which shows the positions of the planets in relation to both Earth and stars at a certain time. When seeking actual evidence of planetary influence, proponents of astrology point to the fact that the Sun and Moon do influence the tides on Earth. They argue that if that is true, why can't they influence human life as well, since water is so crucial to life, and humans are largely composed of water. Powerful logic is not astrology's strong suit.

Horoscopes traditionally place the Earth at the center of the solar system—or even the universe—because that is what was thought to be true when the basic rules of astrology were set down. In astrology, the Sun and Moon are considered "planets" along with the other eight known planets (although there were only five planets known in ancient Babylon and Greece). Astrology contends that the planets hold more influence than other heavenly bodies. Another key element of the astrological scheme is the zodiac, a band of stars that circles the earth and is divided into twelve equal parts—or the familiar *signs*.

THE CONSTELLATIONS OF THE ASTROLOGICAL ZODIAC

Aries (the Ram)

Taurus (the Bull)

Gemini (the Twins)

Cancer (the Crab)

Leo (the Lion)

Virgo (the Maiden)

Libra (the Scales)

Scorpio (the Scorpion)

Sagittarius (the Centaur, also known in astrological history as the Archer)

Capricorn (the Goat, also associated with the god Pan)

Aquarius (the Water Bearer)

Pisces (the Fish)

To see these stars in the sky today, it is sometimes hard to figure how people saw such identifiable pictures in the sky. Yet different civilizations over many centuries often saw the same "pictures."

Each sign has certain characteristics and it is the sign, combined with a particular planet, that determines a person's character. Finally, the

Earth's surface is also divided into twelve parts, called *houses*, each of which also represents certain characteristics. The combination of planets, signs, and houses are the astrological elements that determine character and destiny.

There are many rational objections to the notion that the regular order of the cosmos has some "pull" over events on Earth, other than the force of gravity. In the first place, astrology has traditionally viewed the Earth as the center around which the other planets revolve—an idea pretty well discredited for five hundred years or more. If the five visible planets were once the principal force of astrology, then what did the discovery of previously unknown planets mean to astrological predictions? It is difficult to grant astrology much credence when you take into account the vastness of the universe, the number of stars and the great distances to them, the presence of so many other celestial objects that might pass by unpredictably, and the existence of forces invisible to the human eye such as electromagnetic rays, black holes, and the theoretical "dark matter" that, some astronomers believe, may comprise most of the universe. (For more about these concepts, see part IV.)

In addition, since the time of the ancient Babylonians and Greeks, the constellations have actually moved from where they once were plotted due to the fact that the Earth "wobbles" as it spins on its axis— this is a phenomenon known as the *precession of the equinoxes*, first identified by the Greek astronomer Hipparchus. But, over the centuries, the so-called "fixed stars" have moved in relationship to Earth. In other words, astrologers are working with antiquated data.

Even so, many millions still swear by it. So, this question is simply one more place where the scientist parts ways with the "believer." That's one reason why Shakespeare could dismiss the influence of astrology—as he did in *King Lear* and *Julius Caesar*—yet still call *Romeo and Juliet* "star cross'd lovers" and allow a character in *Hamlet* to remind us, "There are more things in heaven and earth, Horatio, than are dreamt of in your philosophy."

And even the Church fathers, who resisted the pull of astrology, were not above turning to this powerful superstition when it suited their purposes. As historian Daniel J. Boorstin writes, "The great medieval theologians earnestly enlisted the prevailing belief in astrology to reinforce the truths of Christianity. They liked to recall the astrological prediction of the Virgin Birth of Jesus Christ. If Jesus Christ was not Himself subject to the rule of the stars, the stars did give signs of His coming." (*The Discoverers*)

MILESTONES IN THE UNIVERSE:
20 Billion Years Ago–5 B.C.

[Note: Many scholars and historians opt for the dating terms B.C.E. (Before the Common Era) and C.E. (Common Era). In this book, I have chosen to retain the terms B.C. (Before Christ) and A.D. (Anno Domini, or the Year of Our Lord), which are still more familiar to the average reader.]

15–20 Billion Years Ago?   The Big Bang. This is the theoretical Birth of the Universe, the one most widely accepted by modern astronomers. In an instant of cosmic upheaval, all matter, energy, space, and time were created in a cataclysmic burst of heat. Perhaps a billion years after this instant of Creation, gas clouds formed and the first stars began to be born. (The Big Bang is discussed in-depth in part IV.)

5 Billion Years Ago   Our star, the Sun, was born, and the solar system started to come into being. Earth formed about 4.5 billion years ago.

9000–8000 B.C.   The Maya of Central America make astronomical inscriptions and constructions.

A marked bone from this period, or possibly as late as 6500 B.C., found in modern Zaire, is probably used as a record of months and lunar phases.

5508 B.C.   The Year of the Creation (This date was adopted in seventh century A.D. in Constantinople and used by the Eastern Orthodox Church.)

5490 B.C.   The Year of Creation. (This date was reckoned by early Syrian Christians.)

c. 4241 B.C. The Egyptian calendar is devised. It is the first known based on 365 days, or twelve months of thirty days, and five days of festivals. The Egyptian calendar started with the day that Sirius, the Dog Star, rises in line with the Sun in the morning, which coincides with the annual flood of the Nile. (Though this is a widely accepted date, the Egyptian calendar may have been devised as much as 1,500 years earlier.)

4004 B.C. The Year of Creation. (This date was calculated by Irish cleric James Ussher in 1650 A.D. and was considered reliable by most of the European world until geology and biology contradicted it late in the nineteenth century. It is still accepted as "fact" by some "Creationists.")

3760 B.C. The Year of Creation. (This date was recognized in the Hebrew calendar that has been used since the fifteenth century A.D.)

c. 3000 B.C. The Babylonians predict eclipses.

c. 2500 B.C. The Chinese use a vertical pole to project the shadow of the Sun for estimating time.

2296 B.C. Chinese observers record a comet, the earliest known record of a comet sighting.

c. 2200 B.C. The Sumerians use a 360-day year, twelve-month solar calendar along with a 354-day lunar calendar; the calendar has an extra month every eight years to keep it in step with the seasons.

c. 1900 B.C. Stonehenge is erected at sometime in the next three centuries in Bronze Age Britain, possibly as a monumental calculator to chart the movements of the Sun, Moon, and planets.

c. 1750 B.C. Star catalogs and planetary records are compiled in Babylon under Hammurabi.

c. 1600 B.C. The zodiac is identified by Chaldean astrologers in Mesopotamia.

1500 B.C. The gnomon, an L-shaped indicator, is used as a sundial by the Egyptians.

Thutmosis erects in Heliopolis the "Needle of Cleopatra"; its shadow is used to calculate the time, season, and solstices.

c. 1000 B.C.   The Laws of Moses, the beginning of the Hebrew Bible, are first recorded.

c. 850 B.C.   The *Iliad* and the *Odyssey* are written by the blind Greek poet Homer, according to historian Herodotus.

585 B.C.   Thales of Miletus, in what is now Turkey, was said to have correctly predicted a solar eclipse that occurs on May 28, 585. Taking the eclipse to be an ill omen, the Medes and Lydians call off their war.

c. 528 B.C.   Buddhism has its beginnings in India.

c. 520 B.C.   Anaximander of Miletus makes the first known attempt to model the Earth according to scientific principles; his concept is that the Earth is a cylinder with a north-south curvature; he prepares a map of the known Earth based on this idea.

c. 500 B.C.   The Pythagoreans teach that the Earth is a sphere.

c. 480 B.C.   Greek philosopher Oenopides is believed to be the first to calculate the angle that the Earth is tipped with respect to the plane of its orbit; his value of 24 degrees is only half a degree off from the presently accepted value of about 23.5 degrees.

c. 480 B.C.   Pythagorean philosopher Philolaus suggests that there is a central fire around which the Earth, Sun, Moon, and planets revolve; he also believes that the Earth rotates.

c. 400 B.C.   Horoscopes setting out the positions of the planets at the time of an individual's birth become available in Chaldea.

c. 380 B.C.   Democritus, Greek philosopher, recognizes that the Milky Way consists of numerous stars, that the Moon is similar to the Earth, and that matter is composed of atoms.

352 B.C.   Chinese observers report a supernova, the earliest recorded sighting.

c. 350 B.C.   Aristotle, Greek philosopher and tutor of Alexander the Great, believes that the Earth is the center of the universe.

c. 300 B.C.   Chinese astronomers Shih Shen, Gan De, and Wu Xien independently compile star maps that will be used for the next several hundred years.

Also around this time, the Chinese concepts of *yin* and *yang*, or paired opposites, which date to around 2500 B.C., are incorporated into the Chinese conception of the universe.

Elements, a thirteen-volume work by Greek mathematician Euclid, lays out the principles of geometry.

c. 270 B.C.   Aristarchus of Samos (an island near Turkey) challenges Aristotle's teachings by asserting that the Sun is the center of the solar system and that the planets revolve around the Sun; he estimates the distance of the Sun from the Earth by observing the angle between the Sun and the Moon when it is exactly half full.

c. 240 B.C.   Eratosthenes calculates the circumference of the Earth, reaching a figure that is remarkably close to the accepted 24,901.55 miles (40,075.16 kilometers).

165 B.C.   Chinese astronomers record sunspots, probably the first accurately dated record.

46 B.C.   On the advice of Greek astronomer Sosigenes, a calendar of three 365-day years followed by one of 366 days is introduced in Rome under Julius Caesar (the Julian calendar); as a result of changes, to make the seasons correct, the year 46 has 445 days, making it the longest year on record.

44 B.C.   Julius Caesar is assassinated on the Ides of March (March 15). Mount Etna in Sicily erupts, and dust from these eruptions darkens the skies. Roman astronomers report a red comet that is visible in daylight. Its red color is due to the volcanic dust in the air, but Romans believe that the comet is the deified Julius Caesar ascending into heaven.

28 B.C.   Official imperial histories of China begin a sunspot record that continues until 1638 A.D.; the record of 28 B.C. mentions a "black vapor as large as a coin."

7–5 B.C.?   The infant Jesus is born at Bethlehem near Jerusalem, according to Scriptures.

VOICES OF THE UNIVERSE:
The Gospel According to Matthew (2:1–2)

> Now when Jesus was born in Bethlehem of Judea in the days of Herod the King, behold, there came wise men from the east to Jerusalem, Saying, Where is he that is born King of the Jews? for we have seen his star in the east and are come to worship him.

Was the biblical "Star of Bethlehem" an identifiable astronomical event like Halley's Comet?

Few astrological legends have changed the world as profoundly as the story of the Star of Bethlehem. It comes, of course, from the Bible, found in the Gospel According to Matthew. To generations of Christians around the world, this story is probably more familiar than any other in Holy Scripture.

Read each Christmas as part of the birth story of Jesus, it describes the journey of the "wise men" from the East, who are guided by a star. Their number, traditionally thought to be three, is never mentioned in the biblical passage, although the child Jesus is presented with three gifts. They arrive in the city of Jerusalem and enter the court of King Herod, the King of the Jewish people, who rules at the pleasure of the Roman emperor Augustus. These *magi*, or "wise men," tell Herod that they have followed a star to Bethlehem and it is there that the next King of the Jews is to be born. Herod, whom history has judged harshly for his murderous treatment of family and nation, thanks the wise men and sends them packing. He then orders the death of all male infants under two years old in Bethlehem, in an attempt to snuff out any possible threat to this throne.

Scholars and historians have debated many aspects of this story for centuries. Generations of Christians have altered and adapted the tradition of the wise men and the star. But one question has puzzled people for all that time. Could an actual star, or some other extraordinary celestial event, have inspired the Star of Bethlehem?

The best answer is "possibly." But astronomers have argued for centuries about which celestial event is the most likely candidate. It has been called a comet, or a conjunction of planets, or a *supernova* (a rare but extremely bright, short-lived celestial object, the result of an explosion of most of the material in a dying star). Halley's Comet (see page 159) was visible in the year 10 B.C., but that would have been much too early to attach to the birth of Jesus. Some view it as a miracle. Others think it is all part of a myth, written long after the events it described, in order to fit the birth of Jesus into ancient scriptural prophecies.

But in a recent book, *The Star of Bethlehem: The Legacy of the Magi* (Rutgers University Press, 1999), astronomer and physicist Dr. Michael Molnar suggested a novel astronomical candidate for the gospel writer Matthew's star. Molnar proposed that the star was really not a star at all, but the planet Jupiter, seen in the constellation of Aries the Ram. He even offers a date for this appearance: April 17, 6 B.C. That date is consistent with the traditional view that Jesus was born sometime before the death of King Herod in 4 B.C., although some biblical historians argue that Herod actually died in 1 B.C., which would make Dr. Molnar's date less likely.

Molnar's argument is based on several pieces of historical and astronomical evidence. Key among them is the ancient astrological view that Jupiter was a regal star, thus a likely portent of a royal birth. Since it was in Aries, a constellation in the eastern sky, it would have been visible in the East at the time. Molnar dismisses comets, which were at that time generally considered portents of evil or unrest, rather than an omen of a new king. And supernovas were either unknown or ignored by astrologers of the time, according to Dr. Molnar who studied ancient astrological texts of the period. Another key piece of evidence in his argument was his discovery of an ancient Roman coin, which he bought for $50 and which was minted in Syria around A.D. 6. The coin depicts a bright star over the ram, which was also the symbol for ancient Judea. At that time Judea, as the kingdom was called by the Romans, was annexed by Rome.

Of course, the literal truth of this biblical account is like many other events presented in the Bible. Some of them can clearly be documented, others can be linked to specific historical events, such as the fall of Jerusalem. But, ultimately, the Bible is a matter of faith. The Christmas story probably presents the most enduring example of the significance of heavenly portents to the ancient world—as well as to millions of modern Christians.

VOICES OF THE UNIVERSE:
PTOLEMY
(CLAUDIUS PTOLEMAEUS OF ALEXANDRIA)

> Mortal as I am, I know that I am born for a day, but when I follow the serried multitude of the stars in their circular course, my feet no longer touch the earth; I ascend to Zeus himself to feast me on ambrosia, the food of the gods.

Who was Ptolemy and what did he have to do with "one thousand points of light"?

Almost every ancient civilization practiced astrology in some form, but it was the extraordinary life and work of a single man that codified astrological principles as they are still basically practiced today. The lingering importance of astrology in the modern world is not the only contribution made by this second-century Greek scientist named Ptolemy. His ideas about the world, laid out in a book called *Geography*, also introduced the notions of longitude and latitude, and included a world map that had a striking influence on a fellow named Columbus. For better or worse, it is a rather remarkable legacy of a mysterious man who lived nearly two thousand years ago and about whom very little is known.

Most people like to think that, on a clear night, the stars visible in the celestial sphere above are uncountable millions. Maybe billions. Recent discoveries certainly confirm that there are hundreds of billions of galaxies filled with hundreds of billions of stars in the universe. But the idea that the night sky is filled with an uncountable number of stars is a bit of a misconception. When George Bush gave his famed acceptance speech at the Republican National Convention in 1988, in which he spoke of a "thousand points of light," he was closer to the truth. It might seem as though there are millions of points of light up there, but even under perfect conditions, a person with good eyesight can detect about three thousand stars, and the usual number is much lower. The difficulty in counting so many objects makes their number seem overwhelming. Unless you are Ptolemy.

Claudius Ptolemaeus (100–170?) was a Greek astronomer who lived in Egypt during the rule of the Roman emperors Hadrian and Marcus Aurelius. Although little is known of his personal life, we can assume he was patient and meticulous—the ideal librarian. He spent a good portion of his life counting and cataloging the stars. When Ptolemy tallied the stars, he came up with a figure of 1,022 stars visible to his naked eye. (Of course, he was unaided by a telescope, which wouldn't be invented for another 1,500 years.)

When Christopher Columbus took his plan to sail for sailing west to reach the East, one bit of evidence he produced for this scheme was the work of Ptolemy, the thirteen-part work called *Megale syntaxis tes astronomis* ("Great Astronomical Composition"), which included observations that date from A.D. 127 to A.D. 141. (He wrote several other major works after that, so he presumably lived until around A.D. 170) Ptolemy included his catalog of more than 1,000 stars, although some historians argued that Ptolemy had simply filched this list, plagiarizing the one compiled by Hipparchus in 130 B.C.

The work of Ptolemy might have been lost completely, with potentially astonishing consequences for history, except for a remarkable turn of events. Three hundred years after Ptolemy lived and wrote, Rome fell and the "Dark Ages" commenced. At various times over the next few centuries, the library at Alexandria where Ptolemy had worked, was destroyed and attacked variously over the years by Romans and then by early Christian mobs in the fourth century. Ptolemy's system is known only because a single copy of his works had somehow ended up in a monastery in what is now Iran. In 765, shortly after the founding of nearby Baghdad, Arabs made contact with the monastery and its library. There they discovered a treasury of Greek scientific and philosophical material, all of which was eventually translated into Arabic. Fascinated by the Greek achievements in astronomy, the Arabs were so taken with Ptolemy's book that they began to call it *Al Magiste* ("The Greatest"), which was later corrupted to *Almagest*. As Islam spread west over the next few centuries, these Arabic texts also made the trip, until King Alfonso "the Wise," an early Christian king in what had been Muslim Spain, translated the Arabic texts into Latin. Ptolemy's star tables and rules of astrology had made it across the centuries and continents, all the way to Europe.

The book is a description of all that was known in astronomy around Ptolemy's time and compiled everything believed by earlier Greek scholars, such as Aristotle and Pythagoras. Of course, most of these Greek scholars thought that Earth was the center of the universe, and Ptolemy did not contradict them. With geometrical models and tables, the book described the motions of the Sun, Moon, and planets. The most important part of the *Almagest* is its description of what came to be called the "Ptolemaic System," in which the Earth is the center of the universe and the Sun and Moon move around Earth in perfect circles.

Ptolemy's *Almagest* became the standard for astronomy and astrology throughout the later Middle Ages. Astronomy was still based on Plato's principle that all observed motions of heavenly bodies had to be explained in terms of circular motions. Ptolemy's and Aristotle's views became incorporated into church dogma, mainly through the efforts of St. Thomas Aquinas. Supported by the Church, Ptolemy's model of the universe and his astrology, as Carl Sagan notes in *Cosmos*, "helped prevent the advance of astronomy for a millennium."

MILESTONES IN THE UNIVERSE
A.D. 140–A.D. 1642

C. A.D. 140 Greek astronomer Ptolemy writes *Megale Syntaxis tes astronomias*, or *Almagest* ("Great Astronomical Composition," or "The Greatest").

635 The rule that the tail of a comet always points away from the Sun is clearly described in China.

675 The first English sundial is built in Newcastle.

827 A complete translation of Ptolemy's *Megale Syntaxis* is made into Arabic.

832 The "House of Wisdom" is founded in Baghdad; a center of learning, one of many that appeared in the Islamic empire, this one contained an astronomical observatory. Many important Greek documents were translated into Arabic here.

1066 A large comet is sighted; it is connected with the invasion of England by William of Normandy, "the Conqueror," and is depicted in the famed Bayeux Tapestry. Today it is known as Halley's Comet (see page 159).

1271 Marco Polo (1254–1324) begins his legendary great voyage to the Far East, reaching Japan; he returns to Italy in 1295. Although the veracity of Polo's accounts has been called into question, his description of the Orient and its riches and luxuries spur European fascination with reaching the East.

1391 Geoffrey Chaucer, who began writing *The Canterbury Tales* in 1387, writes *A Treatise on the Astrolabe*. In it, he shows how to construct an astrolabe, a device which can be used to compute the position of a star.

1453 The Turks capture Constantinople, forcing many Greek scholars to flee to European cities, bringing vast knowledge with them. This marks a key point in the beginning of the Renaissance.

Johannes Gutenberg prints a Bible in Mainz, Germany, inaugurating the era of movable type, another landmark in the Renaissance.

1492 Columbus reaches the West Indies on his first voyage.

1500 A legendary Chinese scientist, Wan Hu, supposedly ties forty-seven gunpowder rockets to the back of a chair in an effort to build a flying machine; the device explodes, killing Wan Hu who acted as pilot.

1517   Martin Luther posts his *Ninety-five Theses*, inaugurating the Protestant Reformation.

1543   Copernicus's *De Revolutionibus orbium coelestium* ("On the Revolutions of Celestial Bodies") lays out the theory of planetary motions based on the assumption that the Earth and other planets revolve around the Sun.

1558   Elizabeth I becomes queen of England, ushering in a remarkable era of exploration and invention in Great Britain.

1572   Tycho Brahe observes a new star, which he calls a *nova* and which would now be called a *supernova*. As bright as Venus, the star is also recorded by Chinese astronomers and remains visible for fifteen months.

1577   Tycho Brahe tries to determine the distance of the great comet of 1577 from Earth; his crude observations are good enough to demonstrate that the comet has to be at least four times as distant as the Moon.

1583   *Della Causa, principio ed uno* ("On the Cause, the Principle and the Unity"), by Italian priest-philosopher Giordano Bruno, outlines his metaphysical ideas.

1584   Bruno's *Dell Infinito, universo e mondi* ("Of Infinity, the Universe and the World") states that stars form planetary systems and that the universe is infinite.

Bruno's *La Cena della ceneri* ("The Dinner of Ashes") defends the Copernican view of the universe.

1590   Galileo's *De Motu* ("On Motion") refutes Aristotelian physics and relates his experiments with falling bodies.

1592   *Mysterium cosmographicum* ("The Mystery of the Universe") by Johannes Kepler includes the concept that the sphere of each planet is inscribed in or circumscribed about one of the five Platonic regular solids.

1600   Kepler begins assisting Tycho Brahe at his Prague observatory.

Giordano Bruno is accused of heresy because of his adherence to the theory that the Earth revolves around the Sun, and other theories. He is burned at the stake in Rome, on February 17; his connection with the Copernican theory is said to have helped cause Galileo's persecution for writing in favor of a moving Earth.

1602   Brahe's posthumous *Astronomiae instauratae progymnasmata* ("Introduction to the New Astronomy") is published; it contains detailed positions for 777 stars and a description of the 1572 supernova.

1603   In *Uranometria*, German astronomer Johann Bayer introduces the method of describing the locations of stars and of naming them with Greek letters and by the constellation they are in; this system continues to be used today; it is the first attempt at a complete celestial atlas.

Elizabeth I of England dies.

1605   Francis Bacon argues against magic and encourages the development of scientific methods in *Advancement of Learning*.

1607   Jamestown, Virginia, founded; the first permanent English colony in America.

1608   Dutch lens-grinder Hans Lippershey invents the telescope.

1609   Galileo builds his first telescope and, with modifications and improvements, eventually obtains a magnification of about thirty.

Kepler's *Astronomia nova* ("New Astronomy") contains his views that the planets revolve around the Sun in elliptical orbits and that these orbits sweep out equal areas in equal time intervals.

1610   Galileo sees the moons of Jupiter, Saturn's rings, the individual stars of the Milky Way, and the phases of Venus. He reports these discoveries in *Siderius Nuncius* ("Starry Messenger"), which makes him famous across Europe.

1611   Galileo, Thomas Harriot, Johannes Fabricius, and Father Christopher Scheiner all discover sunspots around the same time, though Galileo claims to have seen them four years earlier. He is drawn into a dispute with astronomer Christopher Scheiner, who is also a Jesuit priest, as to who saw them first.

The King James translation of the Bible is published.

1616   Galileo receives a warning from Cardinal Robert Bellarmine that he should not hold or defend the Copernican doctrine that the Earth revolves around the Sun. Copernicus's *De Revolutionibus* is placed on the *Index librorum prohibitorum* of the church, from which it will not be removed until 1835.

1619   Swiss astronomer Johan Cysat discovers the Orion nebula.

Kepler's *Epitome astronomiae Copernicae* ("Epitome of Copernican astronomy"), a defense of the Copernican system, is immediately placed on the *Index* by the Church.

Kepler explains that a comet's tail points away from the Sun as a result of what we now call solar wind—particles from the Sun that push material from the comet out of the head and away from the Sun.

1632   Galileo's *Dialogo sopra i due massimi sistemi del mundo, Tolemaico e Copernico* ("Dialogue Concerning the Two Chief World Systems Ptolemaic and Copernican") puts him in disfavor with the pope, and the book is banned.

1633   The Roman Catholic Inquisition forces Galileo to recant his view that the Earth moves about the Sun.

1639   British astronomer Jeremiah Horrocks observes the Transit of Venus across the Sun, which he had predicted, on November 24.

William Gascoigne invents the micrometer about this time, placing it in the focus of a telescope for measuring the angular distance between stars.

1642   Galileo dies in Arcetri, near Florence, Italy, on January 8.

VOICES OF THE UNIVERSE:
NICHOLAS COPERNICUS (1473–1543)

Finally we shall place the Sun himself at the center of the
Universe. All this is suggested by the systematic procession
of events and the harmony of the whole universe.

Protestant Reformation leader MARTIN LUTHER
on the "upstart astrologer," Nicholas Copernicus:

This fool wishes to reverse the entire science of astronomy.
But sacred scripture tells us that Joshua commanded the
Sun to stand still, and not the Earth.

Why did Martin Luther call Copernicus a "fool"?

The "fool" was a Polish astronomer who developed the remarkable
idea that Earth was a moving planet, an idea for which he is considered the "father of modern astronomy." In 1510, a time when most
others who gave the heavens any thought still accepted the 1,400-year-
old ideas of Aristotle and Ptolemy, Polish scholar Nicholas Copernicus (Mikolaj Kopernik in Polish) had a revolutionary idea about the
universe. But, fearful that his theory would be ridiculed or, worse,
viewed as heresy—he was prophetic on that count—he delayed publication for thirty years.

Born into a wealthy family in Thorn (now Torun, Poland), Copernicus was ten when his father died. He was raised by an uncle, a powerful bishop who saw to it that Copernicus was made a canon, a
church official, guaranteeing the young scholar a decent salary and
few obligations. At eighteen, Copernicus attended the University of
Krakow, studied painting and mathematics, and first developed an
interest in astronomy. In 1497, he went to Italy, at that time the center
of the Renaissance that was transforming European science, arts, government, and culture, and, for the next ten years, he studied astronomy in Bologna, and medicine—still very much a medieval pursuit—

and religious law in Ferrara. When his uncle was appointed bishop of Ermland (East Prussia in north-central Europe), Copernicus was named his physician and personal assistant. And, in 1512, after his uncle's death, he served as a priest, but his true fascination was with mathematics and astronomy, and he turned to the ancient Greeks. Copernicus discovered accounts of Aristarchus of Samos (born about 300 B.C.), a Pythagorean who had been among the first to suggest the notion that the Earth orbits the Sun. Here was the Renaissance concept in a nutshell: the *rebirth* of ancient Greek ideas in sixteenth-century Europe.

Intrigued by the idea dismissed as "absurd opinion" for centuries, Copernicus began to base his calculations on this Sun-centered (*heliocentric*) model. Under the conventionally accepted Earth-centered (*geocentric*) model, which derived from Ptolemy and Aristotle before him, some of the other planets moved in strange ways relative to the Earth. For instance, Mars, Jupiter, and Saturn seemed to move backward (*retrograde*) or in reverse loops. Ptolemy had explained this away with what he called epicycles, smaller circles or loops within the larger circular orbit. Copernicus came to understand that this movement was actually an illusion that occurs because of the differing lengths of the planet's orbits. In his Sun-centered model, Mars, Jupiter, and Saturn were farther away from the Sun, and their orbits were longer than the Earth's orbit. In other words, the Earth "overtook" the other planets as it circled the Sun in its shorter path. Essentially, it is like runners on a track; a runner on an inside lane has a shorter distance than the runners in the outer lanes.

Around 1514, Copernicus gave some friends an essay which described this heliocentric idea. Cautious, he refused to publish it, a decision that was reinforced when a cardinal wrote Copernicus in 1536, pleading with him not to publish the theory.

Nearly thirty years after it was first conceived, his life's work, *De Revolutionibus* "On the Revolutions of the Heavenly Spheres," was finally printed in 1543, with Copernicus on his deathbed. As Copernicus neared death, he had been persuaded to publish by a young Austrian mathematician known as Rheticus (1514–1574), whose real name was Georg Joachim. (He had adopted the pseudonym because his father, with whom he shared the name, was a physician who had been beheaded for sorcery.) The actual publication of Copernicus's work was taken over by Andreas Osiander (1498–1552), a Lutheran minister. Fearing the disapproval of the powerful Reformation leader Martin Luther, who strenuously opposed the idea that the Earth moves, Osiander added a "preface" that was supposedly written by Copernicus. It said, in effect, that the Earth does not move but that his calculations would be easier to understand if one assumed that it did. A few hundred copies were printed a month or so before Copernicus's death and it is not clear whether he saw the published version, although it is said that he was given one on his deathbed but was too weak to hold it.

Written in technical, mathematical language, the book was not a page-turner. Few but the most astute mathematicians could even understand it, and then it was further undermined by the false preface, which seemed to contradict everything Copernicus had set out to prove. The most severe blow to the book came, however, when the Roman Catholic Church placed the book on its index of prohibited books in 1611, where it remained until 1835.

A "fool" to the man who started the great revolution in religion called the Reformation, Copernicus is recognized as one of the giants who began a revolution in science. He is immortalized by a crater on the Moon, named after him. And to mark the five hundredth anniversary of the birth of Copernicus, NASA named an orbiting observatory, operational from 1972 to 1981, after the Polish astronomer who literally turned the universe on its head.

How did a sixteenth-century party animal who lost his nose in a duel change astronomy?

Besides Martin Luther and the Roman Catholic Church, another man who disagreed with Copernicus was Danish astronomer Tycho Brahe (pronounced "brah" or "Brah-hee"). Born three years after Copernicus died, he was one of the most colorful characters in astronomy's history, and spent most of his lifetime developing a systematic approach for observing planets and stars. But, as the telescope had not yet been invented, Brahe used his eyesight and such simple instru-

ments as astrolabes and quadrants—two sticks in a cross-like shape—
to estimate the positions of celestial objects. His observations were far
more precise than those of any earlier astronomers, most of whom
were still relying upon the work of Ptolemy, written 1,400 years earlier. Despite his missing the Copernican boat, the observations of
planetary motion Brahe had made essentially rewrote what was
known of stars and set the stage for all of the major discoverers who
would follow him.

The eldest son of a very wealthy Danish nobleman, Brahe
attended the Lutheran University of Copenhagen where the classical
curriculum included the *trivium* (grammar, rhetoric, and logic) and
the *quadrivium* (geometry, astronomy, arithmetic, and music). Part of
those studies included astrology—the practical side of astronomy,
which then included the study of medicine; it was still widely believed
that the body was influenced by the planets. With a reputation as a bit
of a party guy, Brahe lived the life of a young Danish nobleman. He
didn't keep his nose to the proverbial grindstone. Literally. After arguing with another student over who was the better mathematician,
Brahe and his antagonist decided to settle the matter in a duel in
December 1566. In the course of the duel, Brahe lost a substantial
piece of his nose, which was replaced with a prosthetic made of gold,
silver, and wax.

As a teenaged boy, Tycho saw a partial solar eclipse and became
fascinated with astronomy. On November 11, 1572, he saw what he
thought was a bright new star, what ancient astronomers called a
nova, and which he described in his book *De Stella nova* ("Concerning the New Star"). It was so bright, it could be seen in the daytime.
(What Brahe saw was, in fact, a supernova, which is actually the
explosive death of a large star. For more comprehensive explanations
of novas and supernovas, see part III.) While today this would count
as an unusual sight, in sixteenth-century Europe it was an earthshak-
ing discovery. In the widely accepted view of the time—the Aristotelian idea of an unchanging perfect universe, which was maintained as sacred doctrine by the Church—such a thing as a new star
simply could not be. There was no possibility of change in the heavens beyond the orbit of the Moon because any such change would

suggest that God's Creation was somehow imperfect. What Brahe had witnessed defied conventional wisdom and the doctrine of the Church. Since these were the days in which people were going to war or the dungeons over religion, such an idea was more than a scientific curiosity.

With this discovery and his fame established, Tycho was given a grant of land on the small island of Hven by the Danish king. With his own wealth, and income from tenants on the island, Tycho built an extraordinary scientific observatory and community called Uranisborg (Heaven's Castle) which was equipped with the best astronomical devices then available and a complete staff of observers whose goal was to chart the skies. This wasn't "pure" science though. Brahe, like many of his contemporaries, was convinced of the value of astrology, which was then still deemed to be an important function of astronomy. Possessing more accurate star charts would obviously mean more accurate astrological predictions. Brahe wrote that astrology is "really more reliable than one would think," especially if the positions of the stars were more accurately known.

The Uranisborg observations eventually produced a new catalog of one thousand stars with the positions carefully charted. In 1577, Brahe also observed the path of a bright comet and calculated that it was much farther away then the Moon was, which also upset the conventional wisdom of the day that held that comets appeared between the Moon and Earth. Using his information but still rejecting Copernicus, Brahe devised a model of the universe in which the Sun and Moon revolved around the Earth, but the other planets revolved around the Sun.

Tycho's carousing and his abuse of tenants on his island wore out his welcome and eventually cost him favor with a new Danish king. In 1599, Tycho moved to Prague, where he became imperial mathematician.

His greatest achievement, however, may have come from hiring another astronomer, Johannes Kepler, to join him in Prague in 1600. The two were completely different in temperament and background. Kepler was a mystic and deeply religious thinker at heart; Tycho a carouser and heavy drinker. After one "all-nighter" in 1601, Tycho developed a urinary-tract infection and died eleven days later. But, on his deathbed, he turned over his observations to Kepler and his reported last words were "Let me not seem to have lived in vain."

Mysterium cosmographicum ("Mystery of the Cosmos") by
JOHANNES KEPLER (1596)

> We do not ask for what useful purpose the birds do sing, for song is their pleasure since they were created for singing. Similarly, we ought not to ask why the human mind troubles to fathom the secrets of the heavens. . . . The diversity of the phenomena of Nature is so great, and the treasures hidden in the heavens so rich, precisely in order that the human mind shall never be lacking in fresh nourishment.

Who discovered how the planets move?

Born in Weil (near modern Stuttgart), in south Germany, in 1571, as a boy, Johannes Kepler was sent to the Protestant seminary to prepare for a cleric's life. A child of the Protestant Reformation, he grew up and lived on a frontline battlefield in the wars between Catholics and Protestants. While in school, Kepler met a mathematics teacher who had become a convert to the ideas of Copernicus, and Kepler became fascinated with the Copernican idea of a sun-centered universe. He also detected flaws in that Copernican system, which he attacked with nearly the same kind of fervor that Catholics and Protestants were attacking each other. As he said to his teacher, "I wanted to become a theologian. For a long time I was restless, Now, however, behold how, through my effort, God is being celebrated in astronomy."

To Kepler, God was geometry. In his view, the universe was comprised of the orbits of the planets as spheres stacked inside the five regular solids of classical geometry. Although he was establishing a sacred foundation for his ideas, they would eventually help shake the Church's grip on astronomy.

His family, disgraced when Kepler's father went to fight as a mercenary against the Protestants in the Netherlands, could not afford to keep him in school. Instead he became a mathematics teacher in the small town of Graz, Austria, where he supplemented his income by producing horoscopes. Even a genius has to eat, and, for Kepler, astrology was the chief source of income. In 1998, an astronomer at California's Lick Observatory was checking through some of the observatory's early records when he discovered a very old paper. It turned out to be a horoscope that Kepler had charted for an Austrian nobleman who had been born at 5 P.M. on September 10, 1586.

Kepler was teaching in Graz when *Mysterium cosmigraphicum*, his first defense of Copernicus, was published in 1596. But larger forces were at work in Kepler's world, and the Protestant-versus-Catholic battles then being waged forced the closing of his school in 1598. Kepler was left without work, practically destitute when he received a call from Tycho Brahe to join him in Prague in 1600.

It was a stormy relationship. Kepler, the poor mystic mathematician. Brahe, the wealthy, high-living nobleman. Of his boss, Kepler later said, "Tycho is superlatively rich but knows not how to make use of it." Yet Kepler recognized the value of Brahe's observations of the stars and certainly wanted to share in those discoveries. The first assignment he was given was to plot the path of the planet Mars, which had puzzled astronomers for centuries. To an observer, Mars appears to occasionally move backward in its orbit, in what is known as *retrograde* motion. Since the days of the Greeks, astronomers had been trying to explain this behavior of Mars in a logical, mathematical frame. Kepler was confident he would solve the problem in eight days. It took him eight years. After Brahe's death, Kepler assumed the position of imperial astronomer to the Austrian emperor. In 1606, he wrote *De Stella nova* ("On the New Star"), dealing with the nova of 1604 still known as Kepler's Star. Then, in 1609, he published *Astronomia nova* ("New Astronomy"), which contained two revolutionary new laws of mathematics based on his investigation of Mars. In a declaration that totally shook the way the Earth and its place in the universe were seen, he had concluded that Mars was not moving in a circle—the perfect form, accepted by all thinking people since the time of Pythagoras and Aristotle—but in an *ellipse*, or an oval shape.

KEPLER'S LAWS

1. Every planet follows an oval-shaped path, or *orbit*, around the Sun, called an *ellipse*. The Sun is located at one focus of the elliptical orbit. As a result, the planets are a little closer to the Sun at some times during their orbits.

2. An imaginary line from the center of the Sun to the center of a planet sweeps out the same area in a given time. This means that planets move faster when they are closer to the Sun and slower when they are farthest away.

Ten years later, in *The Harmony of the World*, a grand compilation of all that he had learned and understood, Kepler introduced a third law:

3. The time taken by a planet to make one complete trip around the Sun is its *orbital period*. The square of the period (the period multiplied by itself) divided by the cube of the distance (the distance multiplied twice by itself) is the same for all planets. In simpler terms, a planet that is four times as far from the Sun as another planet takes eight times as long to orbit the Sun.

But these remarkable discoveries did not insulate the astronomer from criticism or the difficult times. He was not safely working away over his astronomical figures, immune from the world. In 1612, he lost his wife and son to the plague, and was excommunicated for his published beliefs. In a world still largely ruled by superstition, he later had to defend his mother from charges of witchcraft. A cantankerous woman, she had been accused of poisoning her neighbors and was in prison, threatened with torture and burning. Six women had already been burned as witches. Kepler was able to save his mother from execution but she was banished from her town, a stark reminder that the world of this mathematician in Reformation Europe was still a world largely ruled by superstition and the harsh authority of the Church.

His mother's safety did not end his troubles. Just as Kepler published the last of his three laws, his world exploded. His adopted hometown of Prague had become ground zero in the holy wars that literally tore Europe apart. Specifically, Kepler's life in Prague came undone with the onset of the Thirty Years' War (1618–1648). Initially a civil war between Protestants and Catholics in the territory ruled by the royals of Austria, the Thirty Years' War eventually spread across the Continent and became a much larger conflict for territory and power. Its roots lay in the hostility between Protestants and Catholics in central Europe (what is today Germany, Austria, parts of Italy, and the Czech Republic). In 1618, the archbishop of Prague ordered a Protestant church destroyed. The local Protestants rebelled in May 1618, and the rebels threw two of the emperor's officials out a window—an incident that became famous as the "Defenestration of Prague."

With the war raging, Kepler was deprived of financial support and was forced to leave Prague. He would remain an imperial astronomer, casting astrological tables for the Duke of Wallenstein until his death in 1630. As Carl Sagan put it, "Kepler stood at a cusp in history; the last scientific astrologer was the first astrophysicist." His contribution to ordering the universe, showing how the planets move in their proper orbits, was a giant leap that laid a foundation for future giants. And in his self-composed epitaph, Kepler wrote:

> *I measured the skies, now the shadows I measure*
> *Skybound was the mind, Earthbound the body rests.*

One of Kepler's other small but significant contributions: in 1610, he coined the word "satellite" to describe the moons of Jupiter.

VOICES OF THE UNIVERSE:
GIORDANO BRUNO,
On the Infinite Universe and Worlds: Fifth Dialogue (1584)

> **Our bodily eye findeth never an end, but is vanquished by the immensity of space . . . There is in the universe neither center nor circumference.**

Who pierced Giordano Bruno's tongue?

For what cause would you allow yourself to spend seven years in prison, be stripped naked, gagged, tied to a stake, paraded through the streets of Rome with your jaw clamped shut, a spike piercing your tongue, and another stuck through the palate? And after that, your prospect is, finally, to be burned alive.

Maybe you would do it for your family. Would you do it for your country? Your religion? Would you do it for an idea?

Four hundred years ago, on February 19, 1600, a brilliant if idiosyncratic thinker—who had taught in the great universities of Paris, Oxford, and Wittenberg—suffered such a fate. It is somewhat curious that his name is not better known among the heroes of human courage. Giordano Bruno was born in Nola, Italy, around 1548. He was educated in the Dominican school in Naples, famed as the place where the great medieval philosopher Thomas Aquinas once taught. In 1565, Bruno entered the Dominican order and was ordained a priest in 1572. But as his education progressed, he found himself increasingly at odds with the Church's views based on Aristotle's logic and its Earth-centered view of the universe.

He left the order in 1576 and began to wander Europe as a traveling philosopher—in his account of scientific development, *The Discoverers*, historian Daniel Boorstin called him an "inspired vagrant"—lecturing in universities in France, where he also began to teach the secret memory skills of the Dominican order. His ideas, based on the principle of association, which are still used in memory-enhancing techniques, were laid out in the book *The Art of Memory*, and made him a bit of a celebrity—a sixteenth-century "Amazing Kreskin."

Bruno's teaching about memory made him the talk of Europe and he literally visited the "crowned heads," winning audiences with King Henry III of France and England's Elizabeth. But chatting up the Queen of England was not a good career choice for Bruno, since she was the head of the Protestant church. When he published a book called *La Cena della ceneri* ("The Dinner of Ashes"), which espoused the new Copernican theory, Bruno sunk deeper into heresy in the eyes of the Church.

Bruno finally went too far for the Church in 1584 with his book, *Dell Infinito, universo e mondi* ("Of Infinity, the Universe, and the World"). Unlike Copernicus before him and Galileo who followed him, Bruno was not an astronomer, but a philosopher. But the ideas of Copernicus intrigued him, and the notion that Earth was not the center of the universe sent him in a different philosophical direction. If the seemingly uncountable stars of the universe were all suns, how many other worlds like Earth might be out there, he wondered? To him, the answer was an "infinite number," filled with the possibility of other beings like ourselves. But notions of infinite other worlds did not sit well with a Church that believed strongly that Earth and mankind were unique among God's handiwork.

Lured back to Rome by the promise of a job, Bruno was denounced and turned over to the Inquisition in 1593, charged with heresy. The Universal Inquisition had been established by Pope Paul III in 1542 to stem the tide of the Protestant Reformation with cruel repression. A council of Dominicans ("dogs of God") tried alleged heretics—trials that often came at the end of extended bouts of excruciating torture. For eight years, Bruno remained in chains in Rome's Castel Sant'Angelo, to be ruthlessly interrogated by the Inquisition. Finally, after a trial by Jesuit Cardinal Robert Bellarmine, one of the greatest intellectuals of the church at that time, Bruno was sentenced to death. He was unrepentant to the end: "I neither ought to recant, nor will I," he told Cardinal Bellarmine. "I have nothing to recant." When the death sentence was pronounced, he defiantly told his accusers, "In pronouncing my sentence, your fear is greater than mine in hearing it." Right after that, the Jesuit and Dominican priests overseeing his trial made sure he wouldn't recant. They clamped his jaw with an iron gag, pierced his tongue with an iron spike, and stuck another spike though his palate. Then on Saturday, February 19, 1600, he was carted through Rome by a cloaked group known as the Company of Mercy and Pity (!). After stripping him of his clothes, the priests burned Bruno at the stake.

Viewed at the time as either a complete heretic or, at best, a misguided philosopher, Bruno left no indelible impact on astronomy or cosmology, save that his visionary notion of other worlds is still considered a possibility by modern astronomers who seek out extraterrestrial life. One of his contemporaries must have certainly taken note of Bruno's fate. But Galileo Galilei probably thought that he, a scientist and mathematician with friends in high places, had very little in common with the eccentric ex-priest who was burned at the stake for his unacceptable ideas.

Bruno's chief piece of astronomical posterity is a crater on the Moon, which has been named for him. The crater is thought to have been created when a meteor smashed into the lunar surface on June 25, 1178, a cosmic crash witnessed and recorded by five British monks.

VOICES OF THE UNIVERSE:
GALILEO GALILEI, *Il Saggiatore* ("The Assayer"), 1623

> Philosophy is written in this grand book—I mean the universe—which stands continually open to our gaze, but it cannot be understood unless one first learns to comprehend the language and interpret the characters in which it is written. It is written in the language of mathematics, and its characters are triangles, circles, and other geometrical figures, without which it is humanly impossible to understand a single word of it; without these, one is wandering about in a dark labyrinth.

Why was Galileo the "Al Gore" of the Renaissance, or, Who *really* invented the telescope?

They always said necessity was the mother of invention. But maybe war is invention's father. The modern electronic computer was largely born of military necessity. In England, during World War II, mathematician Alan Turing was working on the German military code known as Enigma when his team designed Colossus, a precursor to the electronic calculating computer, to help break that code and dramatically change history. About the same time, the Ballistic Research Laboratory in Aberdeen, Maryland, turned to the University of Pennsylvania to create a machine that could quickly calculate trajectories. The result was ENIAC, the first electronic, all-purpose computer, which made its debut in 1946. Even the Internet grew out of Defense Department research.

The past was no different. In the year 1500, around the same time that Leonardo da Vinci was sketching out ideas for wonderful winged devices and helicopters, a Chinese scientist named Wan Hu supposedly tied forty-seven gunpowder rockets to the back of a chair in an effort to build a flying machine. The device exploded, killing Wan— the first death of a test pilot. Weapons design has led to a great many technological leaps.

The telescope, though not exactly an item we think of as a weapon, actually was. In 1609, an Italian mathematician and astronomer heard of a new invention shown at an exposition in Venice. It was an "optical tube" that would bring distant objects close. Its presumed inventor, Hans Lippershey, was trying to patent the idea when the Italian professor got wind of it. When Dutch lens-grinders brought the first "spyglass" to Venice in 1609, the Italian mathematics professor thought he could do better to reinvent the instrument and quickly succeeded. He claimed later to have crafted his version of the spyglass in twenty-four hours. This was a bit of overstatement, kind of like Al Gore's widely reported claim of "fatherhood of the Internet." Galileo's first primitive spyglass was about as powerful as modern binoculars. The name would come later from a friend of Galileo's, who invented it from the Greek for "far-seeing watcher." One of Galileo's first telescopes was given to the doge (the elected chief magistrate) of Venice, and its military and commercial value were immediately apparent. For the first time, ships that were a two-hour sail out of port could be seen.

Like many geniuses, this math professor, Galileo Galilei, recognized the usefulness of such devices and continued to improve on the telescope and other commercially useful inventions. Galileo's stock rose. His salary was doubled and he was assured lifelong tenure. He was appointed court mathematician in Florence and relieved of teaching duties. This good fortune proved to be a double-edged sword. A master of self-promotion, as historian Kitty Ferguson puts it, "Galileo displayed a talent for seeing the unrealized potential of another person's thought or invention and carrying it forward so rapidly and enthusiastically that he was halfway over the horizon before its originator had left the starting line" (*Measuring the Universe*).

But his penchant for success did not win friends and influence people in Galileo's world of Italian academics, religion, and politics in the sixteenth-century. After all, this was the Italy that just a few years earlier, in 1513, had produced *Il Principe* ("The Prince") introducing to the world the concept of "Machiavellian" as a synonym for ambition, power and manipulation. Galileo's good fortune did not sit well with other professors who thought he had taken someone else's idea and run with it. Making powerful enemies was a problem that would plague the great scientist for the rest of his life.

Born near Pisa in 1564, Galileo—like Giordano Bruno in some respects—was apparently not cut out for conventional wisdom. Sent as a boy to a monastery where he was singled out for the priesthood, Galileo and his rather unorthodox father, a struggling musician, had other ideas. The boy was withdrawn from the monastery. Clearly, he was also not cut out for heeding authority. He entered the University of Pisa to study medicine, but became more interested in mathematics and philosophy and left the university without a degree at age seventeen, a "college dropout." Despite the lack of a "sheepskin," Galileo later became a tutor and, eventually, an instructor of mathematics at Pisa, where his fabled Leaning Tower experiment was supposed to have taken place.

The old schoolbook story about the Leaning Tower of Pisa has it that Galileo wanted to disprove Aristotle's belief that the speed of falling objects is proportional to their weight—that is, heavier things fall faster than lighter ones—and demonstrated this to his students by dropping two lead weights or possibly cannonballs simultaneously from the Leaning Tower. This is according to the story told by a sympathetic biographer a few years after Galileo's death. Whether the Tower story was true or was an admirer's exaggeration meant to lionize Galileo, Galileo's views of Aristotle cannot be disputed. He thought the old Greek was "ignorant." However, contradicting Aristotle was not a good career move for those on the sixteenth-century tenure track, and Galileo was dismissed from Pisa. He soon found another position at the University of Padua.

Until this point in scientific history, the basic concepts of astronomy and physics were "debated" by Aristotle's logical methods, as opposed to being tested through investigation and observation—the "scientific method." Galileo thought that the philosophical approach of the Greeks was flawed and believed that precise measurements were more important than formal logic and debating style. The approach, ideas and writings of Copernicus, Brahe, and Kepler showed that Galileo was not alone. In 1597, Galileo read Kepler's *Mysterium cosmographicum*, ("The Mystery of the Universe"), and began to correspond with the mystical mathematician in Graz who shared his belief that Copernicus had it right. Despite his image as one of the daring heroes of science, Galileo wasn't quite ready to go out on that limb. He wrote to Kepler, "I have collected many arguments to refute the [Aristotelian] theory, but I do not publish them for fear of sharing the fate of our master Copernicus. Although he has earned immortal fame with some, with many others (so great is the number of fools) he has become the subject of ridicule and scorn. If there were more people like you, I would publish my speculations. This not being the case, I refrain."

Kepler urged him on, but Galileo was more interested in proof than speculation and he was also apparently more interested in profits than posterity. Those profits eventually came through weaponry—in modern words, a defense contract. Around the time that he was corresponding with Kepler, Galileo produced a "military and geometric compass." Essentially, it was a device fixed to a cannon that was used to measure distance and elevation to fire the cannon more accurately. The compass was a huge success and gave Galileo the financial means that this ambitious son of a poor musician had always sought. It also provoked one of several controversies that would dog Galileo's career. Someone else claimed to have invented the device first and charged Galileo with plagiarism. The charge was disproved—in fact, reversed. But the claim that Galileo had taken someone else's idea and presented it as his own would be one that haunted him again and again.

The telescope was no different. But it provided Galileo with the means to secure his fortune and provide the scientific proof he sought. During the next few months in 1609, Galileo continued to refine and improve his telescope, which he then turned on the sky. In doing so, he made some of the most important discoveries in astronomy. He observed the craters and mountains on the Moon and was struck by the Earth-like appearance of the Moon's landscape. He saw the lunar surface and found it was far from a perfectly smooth sphere, but a landscape of mountains and valleys, pockmarked by craters. He could also see that the dark side of the Moon was faintly illuminated by the Earth, a phenomenon known as "Earthshine." With this information, he proved that the Earth shines as much as the other planets and thus must reflect the light from the Sun.

Directing his telescope on the planets, Galileo discovered that, unlike the stars, which remained like small points in his telescope, the planets appeared like small disks or, as in the case of Venus, as a miniature crescent. This convinced him that the stars are much farther away than the planets. He made landmark discoveries about the other planets, such as the motion of Venus. He began to grasp the dimensions of the Milky Way, discovering that it consists of large numbers of clusters of innumerable stars, and he observed sunspots whose movements demonstrated that the Sun was rotating.

Observing Jupiter, Galileo discovered that it was surrounded by four "stars." (At that time, both the stars as well as the planets were called "stars.") He called these moons, as we now know them to be, "Medicean stars" in honor of the wealthy Medici family, the rulers of Florence and Galileo's powerful patrons. If you lived in Italy in 1600, you wanted the Medicis in your corner. As Galileo observed the motion of these "stars" around Jupiter, he came to the conclusion that they must be circling that planet just as the Moon is circling the Earth. This might seem marginally interesting to most of us today, but to Galileo—and, more significantly, the Church Fathers then trying to hold back the tides of the Protestant Reformation and the Enlightenment with the Inquisition—his discovery was important because it showed that the Earth-Moon system is not unique in God's perfect Creation.

Galileo published a description of what he had seen in March 1610 in a short pamphlet called *Siderius Nuncius* ("The Starry Messenger"). The book made Galileo an overnight sensation, and as Galileo biographer James Reston Jr. wrote, "became the most important book of the seventeenth century" (*Galileo: A Life*).

VOICES OF THE UNIVERSE:
The Starry Messenger, GALILEO

> All the disputes which have tormented philosophers
> through so many ages are exploded at once by the . . . evi-
> dence of our eyes and we are freed from wordy disputes
> upon this subject, for the Galaxy is nothing else but a mass
> of innumerable stars planted together in clusters. Upon
> whatever part of it you direct the telescope, straightway a
> vast crowd of stars presents itself to view.

Although he was moving toward the proof he wanted of what
Copernicus had written, Galileo was still not quite there yet. He was
gathering evidence for what he knew would challenge Church doctrine, and yet Galileo wanted the Church on his side if he could manage that. In fact, there were men in the Church who were willing to
hear him out, a fact made clear when he was invited to Rome for an
audience with the Pope Paul V in 1611 and some of the Church officials actually looked through the telescope he had brought along.
Intrigued, they were still not convinced that Galileo's belief in
Copernicus was warranted.

Why did the Vatican arrest Galileo?

The year 1616 was not a good year for literature; Cervantes and
Shakespeare both died that year. It was also the year that Nicholas
Copernicus's treatise on the solar system, *De Revolutionibus orbium
coelestium* ("On the Revolutions of Celestial Bodies") which had been
published more than seventy years earlier in 1543, was placed on the
Catholic Church's *Index librorum prohibitorum.* You needn't know
much Latin to figure out what that means.

The Church did not think very highly of the Polish astronomer
Copernicus, who said that the Earth moved around the Sun and not
the other way around. This notion contradicted Church dogma of the
time. (The book remained on the banned list until 1835, with the
Church continuing to deny Copernican theory until 1922.)

In that same year of 1616, the fifty-two-year-old Italian astronomer and mathematician Galileo was called on the carpet by one of the cardinals of the Catholic Church, Robert Bellarmine. Galileo was instructed not to continue promoting or defending the forbidden Copernican ideas. This cardinal's warning may have been accompanied by a guided tour of the papal dungeons, where subjects of the Inquisition were introduced to the ingenious techniques that had been devised by the Church Fathers to guarantee a pure faith. Perhaps Galileo could still smell the roasted flesh of Giordano Bruno.

But Galileo had a hard time holding his tongue when it came to his ideas, and, by 1616, he had already put some Vatican noses out of joint. While his improvement on existing telescopes had immediate practical consequences for admirals and generals, Galileo was more interested in setting his sights on the sky.

All of these discoveries confirmed for Galileo that Copernicus was right all along. Perhaps if he had stuck to his telescope, Galileo would have stayed out of trouble. The "old school" Aristotelian professors who scorned his views were still powerful men with the ear of the Church hierarchy. And Galileo was providing plenty of rope for them to hang him. In one open letter, Galileo stated that the Bible should not be taken literally and was irrelevant in scientific arguments. He contended that the burden of proving Copernicus wrong fell on the Church. Summoned to Rome in 1616 by the Jesuit cardinal Robert Bellarmine, Galileo was given his warning, a new set of Papal marching orders, and then spent the next few years in relative quiet, publishing one book, *Il Saggiatore* ("The Assayer"), in 1623.

By 1632, he could stay silent no longer. That year, he published *Dialogo sopra i due massimi sistemi del mundo, Tolemaico e Copernico* ("Dialogue on the Two Chief World Systems, the Ptolemaic and the Copernican"). Unlike most scientific books of the period, it was written in Italian rather than Latin. Ironically, it was also approved by the Church censors. Galileo's fictional dialog featured two men arguing the competing theories; Ptolemaic, in which the Earth was the center, and Copernican, in which the Earth moved around the Sun; a third character was a reasonable man listening to their arguments. In the book, the anti-Copernican character was depicted as a stupid bumbler. One of Galileo's clerical enemies, a Jesuit priest named Christopher Scheiner, convinced Pope Urban VIII that Galileo had modeled this buffoon, Simplicio, on the pope himself. This was a bit of a payback. Scheiner and Galileo were bitter rivals over the question of who had first discovered sunspots. Galileo claimed he had; Scheiner said he had. (In fact, both were beaten to the punch by an Englishman named Thomas Harriot who was mapping the Moon in obscurity, and a Dutchman Johannes Fabricius.) Though a far more scientific man than his predecessors, the pope was not amused.

The temper of the times was against Galileo. The rise of Protestantism had forced the Catholic Church to respond by demonstrating the purity of Church doctrine. Sick in bed, the seventy-year-old Galileo was summoned before the Inquisition in February 1633 for "grave suspicion of heresy." Galileo was told to renounce and halt his teachings of the Copernican view, which he did, presumably reasoning that dead astronomers tell no tales. His book was burned and the sentence against Galileo was publicly read in every university. A sentence of life imprisonment was commuted into a form of house arrest.

According to legend, as Galileo left the Inquisition after recanting his view that the Sun was the center around which the Earth revolved, he muttered under his breath *"E pur se muove."* ("Nevertheless, it moves.") It seems unlikely that Galileo, if he whispered this, would have done so in a voice loud enough to hear. Men were burned for lesser offenses.

The Church officially denied Copernican theory until 1922. (In 1984, a papal commission acknowledged that the Church was wrong about Galileo, but it did not reverse the condemnation of this genius of science until 1992—350 years after his death.)

Did the Church's treatment of Galileo have a "so-called" chilling effect on other scientists and thinkers of the day? Consider the words of Descartes, the noted French philosopher, who wrote in 1634:

"Doubtless you know that Galileo was recently censured by the Inquisitors of the faith, and that his views about the movement of the Earth were condemned as heretical. I must tell you that all the things I explained in my treatise, which included the doctrine of the movement of the Earth, were so interdependent that it is not enough to discover that one of them is false to know that all arguments I was using are unsound. Though I thought they were based on very certain and evident proofs, I would not wish, for anything in the world, to maintain them against the authority of the Church. . . . I desire to live in peace and to continue the life I have begun under the motto *to live well you must live unseen.*"

While living out his sentence, Galileo continued his work, producing a final book in 1638 before going blind, perhaps the result of damage caused by observing the Sun through his telescope. He died on January 8, 1642, in Arcetri, Italy.

On Christmas Day that year, a baby was born in Woolsthorpe, in the countryside of northwest England. He was named Isaac Newton.

VOICES OF THE UNIVERSE:
SIR ISAAC NEWTON (1642–1727)

> I do not know what I may appear to the world; but to myself I seem to have been only like a boy playing on the seashore, and diverting myself now and then finding a smoother pebble or prettier shell than ordinary, whilst the great ocean of truth lay all undiscovered before me.

Did Newton's apple really fall?

What do you call a bachelor said to be a virgin who lived with his college roommate for most of his adult life? You might call him one of the smartest men who ever lived: Isaac Newton of falling-apple fame.

The two most famous apples in western history probably never were. The first, of course, was Eve's apple in the biblical Garden of Eden. A quick reading of Genesis makes it quite clear that the Forbidden Fruit of the Bible was not a tasty McIntosh or Red Delicious. (A fig is actually the more likely candidate for Eve's Forbidden Fruit.)

The second legendary apple was the one that Sir Isaac Newton supposedly saw fall to the ground, inspiring him to figure out the laws of gravity. This story apparently originated with the French writer Voltaire, who had become fascinated with Newton during a stay in England. He had gotten it from Newton's niece. But we don't know where she got it, and it is most likely apocryphal. By the time he was an old man, Newton was repeating the story, but by then it had become the kind of story old men like to tell about their youth, like how far they had to walk in the snow to get to school.

Born prematurely on Christmas Day, 1642, Isaac Newton was a small, sickly child whose father, a yeoman farmer, had died before his birth. It is one of those fascinating "What if"s of history: What if Newton's father had lived? One of the most profound thinkers in history might have simply followed his father as a country farmer, too, and possibly never received an education.

When Newton was three, his mother married Barnabas Smith, the sixty-three-year old rector of North Witham. Newton's mother moved in with Smith, but the young boy remained behind with his maternal grandmother, an abandonment that the young Newton apparently never forgave; in later years, he wrote harshly of his mother and her husband. Seven years later, Rector Smith died and Newton's mother returned home, with three more children. At the age of twelve, Newton was sent to the Free Grammar School in Grantham. They were called grammar schools because they mostly taught Latin grammar. Learning Latin—the international language of European scholars— enabled Newton to read all of the important works of mathematics. He also demonstrated a penchant for the practical, tinkering with strange inventions and acquiring a fascination for sundials.

Newton was, in modern terms, the grammar-school nerd. When there was a powerful storm, he would go outside and attempt to measure its force by jumping against the wind. His schoolmates were predictably not fond of him, and his mother eventually removed him from the school, but Newton clearly wasn't cut out for the life of an English farmer, and the schoolmaster and an uncle convinced Newton's mother to allow him to prepare for university.

Born during the period of the Protectorate, when the English monarchy had been replaced by Cromwell's restrictive Protestantism—no theaters, dancing, or much fun of any sort—Newton was now growing up during the Restoration, when Charles II was returned to the throne under a transformed English monarchy. And at Trinity College in Cambridge, once a home base of Puritanism, monarchists were back in favor. When he arrived there in 1661, Newton was a "sizar," or impoverished scholar, considered the lowest form of university life. Students in his position did duty as valets and servants. They were allowed only leftovers in the dining hall. At eighteen, he was four years older than most other students. But he met a fellow student, John Wickins, and the two became friends who shared rooms for twenty years. Little is known about Wickins except that he did the chores, helped with experiments, and was an excellent copyist.

Although the works of Aristotle were still the prescribed studies, Newton was exposed to the writings of Copernicus, Kepler, and Galileo. He had also been tempted by astrology, buying a book on the subject at a country fair in 1663. In 1665, he received a degree, but Trinity was shut down that year by an outbreak of the bubonic plague. The Great Plague of London claimed more than seventy-five thousand victims—16 percent of the estimated population of London. During the "plague years," Newton returned to Woolsthorpe, his birthplace in the countryside in northeast England, for the next two years.

Apparently without much else to do, Newton essentially rewrote the laws of science. He basically invented calculus. An experiment with a prism led to the discovery of the spectrum and the nature of light. And this was when the legendary apple fell, prompting Newton to ponder why, like an apple, the Moon does not fall to Earth. His answer was the theory of universal gravitation. If the Moon were still, like the apple, it would fall, pulled by Earth's gravity. But the Moon's velocity keeps it in orbit around the Earth. Newton then devised a way to describe this relationship between mass, velocity, and the force of gravity.

But all these concepts remained in his mind for most of the next twenty years. Only following an encounter with astronomer Edmond Halley (1656–1742) did Newton lay out his ideas. As the story goes, astronomer Robert Hooke boasted to the famed architect Christopher Wren and Edmond Halley that he had found laws governing the movement of planets. Wren thought that Hooke was wrong and offered a prize to anyone who could solve the problem. Halley visited Newton and asked what planetary orbits would look like if the planets were attracted to the Sun. Newton replied immediately, "An ellipse, for I have calculated it."

Halley convinced Newton to publish these ideas and sought financial backing from the Royal Society. When the society backed out, Halley underwrote the project himself. In 1687, Newton's work appeared. Commonly known as the *Principia*, it was formally titled *The Mathematical Principles of Natural Philosophy*. In the third volume, Newton coined the word "gravity" (from the Latin word *gravitas*, meaning "heaviness" or "weight") and described the motions of the moons of Jupiter, Saturn, and Earth as well as the movements of the planets around the Sun. Few changes were made to his work for the next two hundred years.

If Halley had done nothing else, his midwifery of *Principia* would have earned him a place in science. But a major astronomer in his own right, Halley achieved astronomical immortality when, in 1705, he predicted that a great comet would return in 1758. Halley died in 1742, but his comet returned as predicted and is known as Halley's Comet (see page 159).

Newton's ideas were so revolutionary and complex that few could understand them, and they were not widely taught for almost another fifty years.

In his later life, the seemingly mild-mannered Newton underwent what some historians believe was a nervous breakdown. Others think he may have ingested mercury during chemistry experiments and suffered from the effects of mercury poisoning. Whatever the cause, his later years were partly devoted to lashing out in bitter attacks at critics and rivals, chief among them Robert Hooke. Newton and Hooke had been at odds for years over discoveries and academic controversies. When *Principia* was published, Hooke even charged Newton with plagiarism. Hooke, perhaps better known for his strides in biology, might have been on the same track as Newton when it came to astronomy and physics, but he never got to the same station. Newton's was not a generous spirit. After Hooke made his charges, Newton had references to Hooke and his ideas struck entirely from later editions of *Principia*. When Hooke died, Newton took his position as president of the Royal Society and ordered a portrait of Hooke destroyed. Newton had similar disagreements with German mathematician Gottfried Wilhelm Leibniz and Royal Astronomer John Flamsteed, whose names were also struck from later editions of *Principia*. He devoted much of his later life to working out a chronology of time that was based on the Bible but went farther back to Greek and Egyptian mythologies. While he was a devout Anglican in public, his chronologies suggested to him that Christianity was simply an offshoot of earlier religious cults.

MILESTONES IN THE UNIVERSE
1644–1687

1644   French philosopher-mathematician René Descartes, in *Principia philosophiae* ("Principles of philosophy"), includes his vortex theory of the origin and state of the solar system, which he perceives in terms of whirling matter.

1647   Johannes Hevelius's *Selenographia* is the first map of the side of the Moon observable from Earth.

1650   Irish Bishop James Ussher sets the date of Creation at 4004 B.C. and the date of Noah's Flood as 2349 B.C., using biblical accounts. His dates are widely accepted for centuries and continue to be accepted by some Christians.

1656   Dutch mathematician-astronomer Christiaan Huygens discovers that the odd "handles" Galileo had seen on Saturn are actually rings; he also discovers Saturn's largest moon, Titan.

1659   Huygens becomes the first to observe surface features on Mars.

1660   The Restoration of the Kingdom of England begins.

1663   The works of Descartes are placed on the Roman Catholic Church's *Index librorum prohibitorum*.

1664   René Descartes's *Le Monde* ("The World"), published posthumously, affirms the Copernican theory; Descartes had abandoned this project after learning of Galileo's problems with the Church.

Robert Hooke discovers the Great Red Spot on Jupiter and Jupiter's rotation.

1665 Giovanni Domenico Cassini (1625–1712) measures the speed of Jupiter's rotation.

1666 Cassini observes the polar ice caps of Mars.

1668 Newton invents the reflecting telescope.

1671 Giovanni Domenico Cassini, who came to Paris in 1669 and took the position of director of the Paris Observatory, discovers Iapetus, a satellite of Jupiter. Cassini also calculates the distance from the Earth to Mars, which enables him to establish the distances of all the planets from the Sun; his calculations are in near agreement with modern measurements.

1672 French physician N. Cassegrain invents the reflecting telescope that is named for him.

Cassini discovers Rhea, a satellite of Saturn.

1675 Cassini discovers that the rings of Saturn are not a single flat disk surrounding the planet; the break in the rings he discovers is still known as the Cassini Division.

England's Greenwich Observatory is founded by King Charles II.

1679 Cassini publishes *Carte de la lune*, a map of the Moon.

Catalogus stellarum australium by Edmond Halley gives the locations and descriptions of 341 southern stars, the first time that stars observable from south of the equator have been cataloged.

1682 Edmond Halley observes the "great comet" which will be named for him in 1705, after he correctly predicts that the comet will return in 1758.

1684 Cassini discovers Dione and Thetys, satellites of Saturn.

1686 Newton presents the manuscript of *De Motu corporu* ("The Motion of Bodies"), the first volume of his *Principia*.

1687 Newton presents *Philosophiae naturalis principia mathematica* ("The Mathematical Principles of Natural Philosophy"), better known as the *Principia*, which establishes his Three Laws of Motion and the Law of Universal Gravitation.

A thread that began in ancient Greece some 2,500 years ago with those obscure Greek names—Thales, Hipparchus, Aristarchus—stretched over the centuries to the seventeenth-century England of Newton. Each scientist added to the intricate web spun by those who came before. As Newton himself wrote to fellow astronomer Robert Hooke, "If I have seen further, it is by standing on the shoulders of giants." (Ironically, he and Hooke later had a falling-out in Newton's cranky older days, and Newton spent considerable energy making sure Hooke was miserable, by keeping him out of science societies.)

But using the terms of biblical genealogy, we can say that Aristarchus begat Ptolemy, Ptolemy begat Copernicus, Copernicus begat Brahe, Brahe begat Kepler, Kepler begat Galileo, Galileo begat Newton.

Well beyond the scientific revolution these men spawned—the early Greeks, and such rabble-rousers as Giordano Bruno and Galileo, and, finally, a scientific genius like Newton, can also be credited with helping spark other revolutions. The liberating quality of reason helped science break free of the Church's hold. It was only a matter of time before political philosophy followed suit. In a sense, a clear line can be drawn from Newton's natural laws to the democratic philosophy that was taking shape in England at the time. Newton's ideas, which helped shake science free of some of the restraints of superstition and religious orthodoxy, inspired men of the Enlightenment, like political philosopher John Locke (1632–1704), who wrote in The Inalienable Rights of Man, "A government is not free to do as it pleases. The law of nature, as revealed by Newton, stands as an eternal rule to all men." Locke's ideas became one of the key foundations of Thomas Jefferson's Declaration of Independence.

These are the threads of history, interweaving scientists and philosophers—from the ancient world that conceived the democratic idea—through the centuries, thick with superstition and the inability to question authority, finally to a time in which thinkers who dared to question reshaped both the human view of the universe and the world in which we live.

PART II

ACROSS THE GULF

Give me the splendid silent sun with
all his beams full-dazzling.

WALT WHITMAN
Give Me the Splendid Silent Sun, 1865

Lo, the moon ascending,
Up from the east, the silvery round moon,
Beautiful over the house-tops, ghastly phantom moon,
Immense and silent moon.

WALT WHITMAN
Song of the Universal, 1881

Space—the final frontier . . . These are the voyages of the starship *Enterprise.* Its five-year mission: to explore strange new
worlds, to seek out new life and new civilizations, to boldly go
where no man has gone before.

GENE RODDENBERRY, *Star Trek,*1966–69

What is space?

Is anybody in charge of space?

How and when was the solar system created?

How big is the solar system?

What is a planet?

How do you tell a planet from a star?

How big is the Sun?

What sort of star is the Sun?

Why are there spots on the Sun?

Does the Sun make noise?

What's the difference between a solar eclipse and a lunar eclipse?

What are the northern lights?

Will the Sun always come out?

Why is it difficult to see Mercury?

Why is Venus so hot?

If Venus is a planet, why is it called the "Evening Star"?

What is the transit of Venus?

Why is there life on Earth?

Does Earth wobble?

How long is the Moon's month: 27 days or 29 days?

What's up with the Man in the Moon?

Where did the Moon come from?

Is a "blue moon" really blue?

Who were the lunatics?

Is the Moon Earth's only satellite?

Is Mars really red?

Who dug those Martian canals?

Mars: Desert or wetland?

Mars: Dead or alive?

Will we ever go to Mars?

Did an asteroid kill the dinosaurs?

What is Jupiter's Great Red Spot?

Could Jupiter be an underachieving star at the center of a solar system that fizzled?

What are Saturn's rings?

Why isn't Uranus called George?

How did they find Neptune without a telescope?

Who found Planet X?

Will we ever visit Pluto?

"Shooting stars" and comets: What's the difference?

Where do meteoroids come from?

What is a comet?

Who was Halley?

What happened to Comet Shoemaker-Levy 9?

What would happen if a comet hit the Earth?

If meteors come from comets, where does a meteorite come from?

Does a meteorite carry no-fault insurance?

Are there any other planets in our solar system?

In the Beginning . . . there was gas and dust.

Not very impressive sounding, is it? And it certainly wouldn't make much of an opening for a divinely inspired book.

Gas and dust. To some of us, these words conjure up getting off the highway to make a "pit stop" only to find a beat-up filling station with a filthy rest room. Now there's a cosmic thought!

But gas and dust are the basic ingredients that, scientists tell us, contributed to the beginnings of the universe and the solar system in which we live.

There are many ways to study and understand the universe. This book puts us—the tiny but unique piece of rock called Earth—at the center ring of the cosmic circus. The story starts, of course, some 10 to 20 billion years ago with the Big Bang, the theoretical beginning of matter, energy, space, and time. Working its way forward, it finally gets around to this rather insignificant speck called Earth, a smallish lump in a solar system in an average galaxy.

This section of *Don't Know Much About the Universe* begins with a look at our own little neighborhood—or celestial backyard—the solar system that revolves around our star, the Sun. The solar system in which Earth resides consists of the Sun—although often described as a rather typical star, it is special in many respects—and all the bodies orbiting it: nine (although some have begun to say eight) planets, their moons, the asteroids, and the comets. While all of that sounds like an impressive amount of stuff, the Sun actually contains 99.86 percent of the entire mass of the solar system. If you didn't feel insignificant before, get used to it now. This book will make you realize what a tiny mote in space we, a few humans living on Earth, actually are.

We exist in a rather small corner of the universe—one small planet orbiting our star alongside billions of other stars tucked inside an ordinary galaxy moving through the vast universe filled with hundreds of billions of other galaxies—all of it moving through space.

What is space?

"Space," as Captain Kirk of the original *Star Trek* series told us at the beginning of each episode, "is the final frontier."

That may be true, but what is space? To put it in simple dictionary terms, "space" is "The expanse in which the solar system, stars, and galaxies exist; the universe," according to *American Heritage*. In another variation, "space" is "the region of this expanse beyond Earth's atmosphere." The English word "space" comes from Middle English for "area," from Old French *espace*, from the Latin *spatium*.

To put it simply, space is everything that is "Out There." Of course, we earthlings might like to think of space as "out there." But, in fact, our smallish chunk of rock called Earth is also "lost in space." It travels on its regularly scheduled turn around the Sun, even as our solar system moves through the Milky Way galaxy, which is itself moving through the universe, which is also on the move, expanding in all directions for nobody knows how far. This idea is what causes such great consternation to Woody Allen's five-year-old alter ego in the classic *Annie Hall*. Taken to the doctor by his impatient Jewish mother, the worried—neurotic is probably more accurate—boy tells the doctor he has heard the universe is going to expand and then collapse. Smoking a cigarette and blowing a thick plume of smoke into the boy's face, the doctor laughs him off and tells him it won't happen for billions of years. The doctor may be right, but it is all rather mind-boggling, isn't it?

As we go hurtling through the seemingly infinite blackness of the universe—what astronomers, cosmologists, and physicists now call "space time"—all that separates Earth from space is the wafer-thin blanket called "the atmosphere." Perhaps "amniotic sac" is more appropriate. Instead of a life-giving fluid, however, this delicate membrane surrounding our planet is made of a variety of gases that supply us with the air and water that make life possible, keep out harmful radiation, and help to maintain the planet's climate, which, in turn, allows life to continue.

To all Parents of Children Who Ask That Eternal Question "Why is the sky blue?," here goes: *This atmosphere is what makes the sky blue*. The minute particles of matter and air molecules in Earth's

atmosphere intercept the Sun's white light, and act like a prism, sepa-
rating the light waves, each of which we see as a different color. The
shorter (blue) wavelengths are more visible than the longer (red)
wavelengths. And that is why the sky looks blue. And that's where
those pink sunrises and red sunsets come in. At those times of day, the
angle of the sunlight changes. The light travels farther through the air,
and the blue wavelengths are scattered, making the red wavelengths
more visible.

Like Linus's security blanket in *Peanuts*, the atmosphere is a
remarkably thin layer of protection, considering the vastness of the
solar system and the galaxies beyond. By studying deep-ice core samples and fossilized tree resin—the amber that inspired *Jurassic Park*—
scientists have discovered that the atmosphere has been the way it is
for the last hundred million years or so, apart from some relatively
slight changes that may account for Ice Ages and dinosaur extinctions.
Composed of about 78 percent nitrogen, 20 percent oxygen, with a bit
of argon and carbon dioxide, and traces of hydrogen, neon, helium,
krypton, xenon, methane, and ozone, the Earth's atmosphere is a
remarkable combination that has allowed life to flourish. If we didn't
mess with it, that atmosphere might presumably remain as it is for a
few million years more. That is, unless we continue to fool with
Mother Nature. In the view of a growing number of scientists, all of
the gases released from burning fossil fuels and cutting down forests
during the past two hundred–odd years since the Industrial Revolution began to threaten that balance. There is a real fear of the havoc
being wreaked upon the tender tissue that surrounds us and allows life
on Earth to go on, protecting us from the cold, airless eternity of the
universe. It comes in layers, five of them: *troposphere*, *stratosphere*,
mesosphere, *thermosphere*, and *exosphere*.

Starting from Earth's surface and working out, the closest layer is
the one in which we live, the troposphere; it is the only layer in which
humans can survive without protection. Warmed by the Sun, whose
heat is stored in the ground and water, the troposphere reaches up to
an average of about 7.5 miles (11 kilometers) above Earth's surface.
(The troposphere is slightly thicker at the equator and thinner at the
poles.) As you rise through the troposphere, the temperature falls,
cooling at a rate of about 10°F for every 3,000 feet (900 meters). As

you climb through this lower portion of the atmosphere, the level of oxygen also decreases, and there is a marked increase in ultraviolet (UV) radiation. This is why high-altitude climbers must carry oxygen tanks and plenty of sun lotion. As the popular song recently put it, "Trust me on the sunblock."

The second layer of atmosphere is the stratosphere, which begins about 6 miles (10 kilometers) above Earth's surface in the polar regions and about 10 miles (16 kilometers) near the equator. The air is noticeably thinner and, as we climb higher, it starts getting hotter, as the exposure to the full force of the Sun increases. The nearly cloudless and very dry stratosphere, filled with fast jet streams, is where pilots like to cruise, and airplanes make the best time. It is also in the stratosphere that the much-discussed *ozone* layer is found, a band of gas that is not breathable but is still crucial to life on Earth. A form of oxygen, ozone provides the lifesaving screen that blocks out much of the harmful ultraviolet (UV) light from the sun, UV radiation that would endanger life on Earth if it reached the Earth's surface in greater quantities. The detection of a hole in the ozone layer about a decade ago raised extreme worries about the increasing amounts of UV radiation penetrating to the Earth's surface. One immediate result: an increasing number of skin cancers.

The third layer, beginning at an altitude of about 30 miles (48 kilometers) and extending to about 50 miles (80 kilometers), is the mesosphere, and it continues to about 55 miles (85 kilometers) above the Earth. While the atmosphere is far too thin to breathe here, it is still an important part of Earth's life-support system. The mesosphere acts as Earth's "force field," a shield against space debris. Dust from comet tails and meteoroids approaching the Earth burn up in the mesosphere. These bits of space debris, as much as 50,000 tons of it annually, speed along at 44,000 miles (71,000 kilometers) per second, are heated by friction, and rendered harmless, scorched into cinders. (A few bits of this speeding space litter do manage to find their way through the mesosphere and down to *terra firma*. More about that later in this section.)

At an altitude of about 53 miles (85 kilometers), the thermosphere begins and continues about 300 miles (480 kilometers) into space. Completely exposed to the Sun's radiation, the thermosphere has only a tiny fraction of the gases that are in the atmosphere. Although it is very thin, there is still enough "air," mostly oxygen, in this layer to capture the Sun's heat. Temperatures in the thermosphere soar from about –135°F (–93°C) at an altitude of 55 miles (89 kilometers) to over 2700°F (1500°C) in the thermopause, the upper area of the thermosphere.

Finally, comes the ultimate layer, the exosphere, which begins at the thermopause and eventually merges with the solar wind. Composed mostly of helium and hydrogen, it reaches out for approximately 625 miles (1,000 kilometers) from the surface of Earth, gradually diminishing until the end of the atmosphere has been reached. The exosphere has so little air that satellites and spacecraft orbiting Earth in the region encounter almost no resistance, and the atoms and molecules of the air here travel so fast that they overcome the force of Earth's gravity and escape into space. The bad news is that Earth is slowly losing its atmosphere. The good news, however, is that this process will take billions of years.

Welcome to space!

Is anybody in charge of space?

Until the federation of *Star Trek* fame actually comes along, or the empire from *Star Wars* gets the upper hand, space is basically shared by all of us. In 1967, the United Nations Outer Space Treaty was signed, providing guidelines for the peaceful exploration and exploitation of space, the moon, and other celestial bodies. It is based on a humanist notion that space belongs to everyone and that all nations can explore and use it. It has been ratified by ninety-four states and signed by an additional twenty-seven as of late 1999.

The Third United Nations Conference on the Peaceful Uses of Outer Space (UNISPACE III) was held in Vienna in July 1999. With the end of the Cold War and a shift in emphasis on space as a commercial venture instead of a potential battleground for dueling ballistic missiles, the focus of UNISPACE is to create a blueprint for practical, peaceful exploration of outer space. The 1999 Vienna meeting outlined broad areas for peaceful uses of space in the twenty-first century.

They include protecting Earth's environment and marine resources; using space applications for human security, development, and welfare; advancing scientific knowledge of space and safeguarding the space environment; enhancing education and training opportunities; and ensuring public awareness of the importance of space activities.

Space as an international issue is on the front burner because of the plan for a space-based missile defense program being debated in the United States. At the height of the Cold War years, when American and Soviet military plans were based on nuclear deterrence, the only protection from a missile launch was the fear of Mutually Assured Destruction, or MAD, in which both sides finally realized that an all-out nuclear war was an unwinnable proposition. A series of treaties during the 1970s and 1980s gradually reduced the possibility of such an all-out war. But, in 1983, President Ronald Reagan believed that there was a better solution, a space-based defensive shield that could destroy incoming missiles. Known as the Strategic Defense Initiative (SDI), it was derisively nicknamed the "Star Wars" defense by skeptical critics, but the name stuck. The Pentagon spent heavily on the unproved technology, which provoked the Soviet Union to attempt to keep up. That expensive arms race is credited by many historians to have accelerated, in part, the collapse of the corrupt and inefficient Soviet economy, bringing about the end of Soviet Communism and, with it, the end of the Cold War. (Not everyone agrees with that particular historical assessment, including Mikhail Gorbachev, the architect of the Soviet Union's collapse. In 2001, Gorbachev said Reagan's arms buildup played no role in the Soviet Union's downfall. "The Soviet Union was a victim of battles within the country itself," Gorbachev told Newsweek magazine. "People were not free, they were unhappy, and this couldn't be ignored. The arms race was not decisive.")

But with the Soviet threat reduced, many military planners still see a potential threat from missiles launched by smaller, "rogue" nations or even terrorists who have purchased old, Soviet-era technology. And the idea of the SDI, or "Star Wars Defense," has not been abandoned. Nearly $100 billion has been spent on developing the technology for a space-based missile defense system, which, critics also argue, would break certain nuclear control treaties signed earlier. Upon taking office in 2001, the Bush administration has stated that development of the antimissile shield is still one of its military priorities. The fate of this technology, which sharply divides the scientific community, is undecided. As the editors of *Scientific American* wrote in a June 2001 editorial:

> Regarding strategic missile defense, researchers' best guess is that a reliable system is infeasible. . . . Until [proponents of missile defense] can provide solid evidence that a system would work against plausible countermeasures, any discussions of committing to building one—let alone meeting a detailed timeline—is premature. It is one thing for a software company to hype a product and then fail to deliver; it is another when the failure concerns nuclear weapons, for which "vaporware" takes on a whole new, literal meaning.

Continuing development, against enormous technological odds, will prove extremely costly. And the cost to the United States in its relationships with other countries is also questioned by skeptics who feel that America's allies will be more exposed to an attack by someone attempting to beat the completion of the defense shield. A third argument against development: no defense in history has ever been foolproof. From the Great Wall of China to the Maginot Line, history shows that new weapons are destined to be created to defeat or diminish the most brilliant defense. Finally, there are many scientists who feel that space should remain a "demilitarized zone," free from all weapons. Although that seems a quixotic hope at best, given mankind's history, it is what the UN treaty calls for.

Besides the end of the Cold War–era "space race" between the United States and the Soviet Union, the greatest difference between past concerns and present space issues is the presence of private enterprise at the conference table when space ventures are now discussed. Space is going to be Big Business. Many commercial ventures are being formed to explore and exploit space. From manufacturing in zero gravity to space tourism, a new era of space commerce is upon us. Who will be the first corporate sponsor of a shuttle launch or a Mars lander? It gives the phrase "advertising space" a whole new meaning.

How and when was the solar system created?

"This just in: Scientists observing the void have just witnessed a tremendous explosion. Within fractions of seconds, everything in the universe was created."

It would be so easy if CNN had been there. If there had been video cameras around at the moment we call the Big Bang, the theoretical instant in which all energy and matter were created. No more debates about Creation, Genesis, God, or any of the other hundreds of mythic, religious, and philosophical explanations for the beginning of the universe. But, obviously, there is no such news report. The best explanation for the history of the solar system—and the universe—has been pieced together from bits of evidence that have been collected, mostly, during the past thirty years.

The solar system in which Earth is found is thought to have begun life as a great cloud of gas and dust that began to collapse in on itself, under the force of its own gravity, about 4.6 billion years ago. As the cloud shrank, it started to spin and flatten out into a disk. Eventually, the center began to heat up as the particles of gas and dust smashed into each other, forming the star we know as the Sun. The rest of the material went on to form the planets, moons, asteroids, comets, and other debris that make up the solar system.

Once the solar system seemed a simple place, its members forming a neat hierarchy: a central star, nine planets in regular orbits, a dozen or so dead moons, the occasional comet, and a collection of asteroids. During the past thirty years, that view has been transformed. Moons can be larger than planets, and comets whiz through by the trillions. So here it is. Meet your solar system.

SOLAR SYSTEM: VITAL STATISTICS

Age: 4.6 billion years

Known planets: 9, but even that's debatable

Known life-bearing planets: 1

Satellites: at least 63, but we keep finding new ones all the time

Largest planet: Jupiter (318 times the mass of Earth)

Smallest planet: Pluto (0.2 the diameter of Earth)

How big is the solar system?

Once you begin to measure distance outside of a relatively small area around the Sun, using earthly measurements such as miles (or kilometers) becomes pointless because the numbers increase in size until they become unmanageable. Astronomers measure distance within the solar system using the term *astronomical unit* (AU). One astronomical unit is the average distance between the Earth and the Sun, which is about 93 million miles (150 million kilometers).

So, for some perspective: The distance between the Sun and the planet Jupiter averages about 5 AU (465 million miles; 750 million kilometers). Pluto, the most distant planet, is about 39 AU from the Sun, or 3.65 billion miles (5.9 billion kilometers). If we consider that the solar system includes everything within the orbit of Pluto, that area expands geometrically to an impossibly large number.

But the solar system doesn't even end with Pluto. The solar system's most-distant known object, at present, is a body called 1996 TL66, which lies some eighty times as far from the Sun as Earth is, and twice the distance of Pluto.

What is a planet?

This used to be fairly easy to answer: Any of the nine large objects that orbit the Sun. Or, in a wider sense, any large celestial body in orbit around a star. But this is one of those many instances where things have gotten a bit tricky in recent years.

"They call me the wanderer," an old rock 'n' roll song says. "I roam around, around, around." Sort of like the planets, going round and round the Sun, in regular orbits that have been observed and recorded for centuries. That's what the ancient Greeks called them. The word "planet" comes from the ancient Greek word for "wanderer," because the planets seemed to move against the more fixed lights of the stars.

Putting it less poetically, a planet is a large celestial body in orbit around a star, composed of rock, metal, or gas. They do not produce light, but reflect the light of their parent star—the Sun, in the case of the planets in our own solar system.

The planets' regular movements are described in two basic ways: *revolution* and *rotation*. Each planet orbits around the Sun, or *revolves*. Each of these revolutions equals one planetary year—365 days in the case of the Earth. Each planet also turns constantly, like spinning tops, and this is called *rotation*. A single rotation equals one of that planet's days—again, in the example of the Earth, a rotation takes approximately twenty-four hours.

Since the discovery of Pluto in 1930, astronomers have taught that there are nine planets in our solar system. The four planets closest to the Sun are comparatively small and rocky and are called the inner, or "terrestrial" planets:

Mercury

Venus

Earth

Mars

The outer planets—with the exception of Pluto—are called the major planets, "jovian planets," or "gas giants."

Jupiter

Saturn

Uranus

Neptune

Pluto

Pluto is a celestial oddball, which has more in common with the moons of the outer planets, and is neither terrestrial nor jovian. Some astronomers, in what is a polite fight over nomenclature, have begun to wonder whether it should even be rightfully called a planet. In fact, if you visit the Rose Center for Earth and Space in New York City, one of the country's most prominent planetariums, you will find that Pluto's status is decidedly ambiguous. The Rose Center demoted Pluto from the ranks of the nine, and it has been reassigned to being just one of more than three hundred icy bodies orbiting beyond Neptune in a region of the solar system called the Kuiper (pronounced KY-per) Belt (see page 157). While most mainstream astronomers disagree with the Rose Center's decision, it does point up the fact that science "facts" must sometimes change to suit new discoveries.

Over the years, several mnemonics have been devised to help keep the nine planets in order: **M**any **V**ery **E**arly **M**ornings, **J**ack **S**tood **U**nder **N**ancy's **P**orch.

Another favorite: **M**y **V**ery **E**ducated **M**other **J**ust **S**erved **U**s **N**ine **P**izzas. (This one does have the advantage of using the number nine.)

How do you tell a planet from a star?

Here's a clue. "Twinkle, twinkle, little star." Stars do appear to twinkle when seen from the Earth. When the light of distant stars passes through the layers of Earth's atmosphere, the temperature differences in the atmospheric layers create a shimmery effect, much in the same way that a hot road seems to shimmer in the summertime. Unlike stars, planets do not "twinkle" in the night sky but are visible as steady and usually bright lights. But aside from appearance, what is the real difference between them?

Simply put, planets are dark bodies that revolve around stars. They do not produce light, as stars do, but only reflect the light from the nearby star—in the case of our solar system, the Sun.

Most of the lights we can see at night, whether faint or bright, are burning balls of gas and plasma situated in the heavens at extraordinary distances from the Earth. We call those seemingly countless little pinpricks of light "stars," but a few of them are planets within our solar system. Known from ancient times, five of these planets (Mercury, Venus, Mars, Jupiter, and Saturn) revealed their true nature by moving slowly from night to night against the background of the fixed stars. Together with the Sun, Moon, and Earth, these five were familiar throughout recorded history, and have been part of human myth and religion since the earliest times.

The other three planets went unnoticed for a long time because, for the most part, they can only be seen with the assistance of telescopes. Uranus was discovered in 1781. (Though extremely faint, Uranus can be seen unaided under ideal conditions.) Neptune was discovered in 1846, and, finally, in 1930, Pluto's presence was revealed through the study of photographs of the night sky.

MILESTONES IN THE UNIVERSE
1703–1804

1703   Newton is elected president of Royal Society.

1735   Francesco Algarotti writes a simplified version of Newton's optics titled *Newtonianismo per le dame* ("Newtonianism for the Ladies"); it becomes one of the most popular explanations of Newton's physics.

Charles Marie de la Condamine leads an expedition to Peru to measure the curvature of the Earth at the equator and sends back samples of rubber and curare.

1743   Benjamin Franklin establishes the American Philosophical Society in Philadelphia, the first scientific society in the United States, around the same time he invents the Franklin stove.

1745   Comte Georges-Louis Leclerc de Buffon proposes that the Earth was formed when a comet collided with the Sun.

1746   Harvard University establishes the first laboratory for experimental physics under John Winthrop.

1749   Gabrielle Emilie le Tonnelier de Breteuil, Marquise du Châtelet, completes the only French translation of Newton's *Prin-*

cipia ever made into French. She dies in childbirth later that year. Her lover, Voltaire, has encouraged her in this work and writes the preface for the 1759 first edition.

1751 Diderot and Jean Le Rond d'Alembert publish the first volume of a seventeen-volume *Encyclopedia*, a rational dictionary of science and art.

1752 In June, Benjamin Franklin performs his famous kite experiment which shows that lightning is a form of electricity. The following year, George Wilhelm Richmann is killed while performing a similar experiment. In 1760, Franklin begins placing lightning rods in Philadelphia.

Great Britain and the American colonies adopt the Gregorian calendar by having September 14 directly follow September 3.

1755 In *Allgemeine Naturgeschichte und theorie des himmels* ("General Natural History and Theory of the Heavens"), German philosopher Immanuel Kant (1724–1804) suggests that the observed nebulas are large star systems, like the Milky Way, and that the solar system originated from a dust cloud.

1759 John Harrison completes Number Four, the marine chronometer that eventually wins the British Board of Longitude's prize for a practical way to find longitude at sea. In 1765, he is granted half the prize, but jealous competitors cause the second half of the prize to be withheld for twenty years.

1761 A worldwide effort to observe the Transit of Venus is made. The information will be used to calculate the distance from the Earth to the Sun.

Observing the Transit of Venus, Russian Scientist Mikhail Lomonosov discovers that Venus has an atmosphere.

1768 Publication of *Encyclopedia Britannica* begins in weekly installments. The first bound edition in three volumes appears in 1771.

1769 Captain James Cook's voyage to the South Pacific proves that there is no large southern continent except Australia.

1779   Comte de Buffon argues in *Epoques de la nature* ("Epochs of Nature") that 75,000 years have elapsed since Creation; this is the first modern speculation that the Earth is older than biblical evidence shows, which is about 6,000 years.

1781   William Herschel (1738–1822) discovers the planet Uranus on March 13, although he first believes it to be a comet.

1783   Following a deadly volcanic eruption in Iceland, which kills one fifth of the population, Benjamin Franklin speculates that dust and gases from the volcano could lower temperatures by screening out the radiation from the Sun.

The brothers Montgolfier of France demonstrate the first successful hot-air balloon with a flight over Paris on *June 5, 1783.*

1801   Giuseppe Piazzi discovers the first asteroid, Ceres.

1802   Heinrich Wilhelm Olbers (1758–1840) discovers the second asteroid, Pallas.

1804   Karl Ludwig Harding discovers the third asteroid, Juno, and Olbers discovers the fourth, Vesta.

VOICES OF THE UNIVERSE:
The Walrus and the Carpenter, LEWIS CARROLL

The Sun was shining on the sea,
Shining with all his might;
He did his very best to make
The billows smooth and bright—
And this was odd, because it was
The middle of the night.

The moon was shining sulkily,
Because she thought the sun
Had got no business to be there
After the day was done—
"It's very rude of him," she said,
"To come and spoil the fun."

THE SUN

If you went to the Sun

Sunlight streaming through the window—warming, pleasant. From the beginning of human times, the Sun's role as giver of life has been apparent. In almost every culture, the Sun has been honored and given a central role in its mythology. The Egyptians developed one of the most elaborate solar cults five thousand years ago, with the pharaoh viewed as an incarnation of the Sun God. To the Greeks, the Sun God Apollo, whose golden chariot carried the Sun across the sky, was the god of light and possessed great moral and intellectual status. A grimmer mythology grew among the Aztecs of Mesoamerica who believed that they were the "Sun's Chosen People." To ensure that the Sun would return each day, the Aztec emperor kept a mythic pact with the sun gods, which required copious amounts of "precious water," the blood of sacrificial victims. One of these Aztec sun gods was Tezcatlipoca, and, in the Aztec capital, a handsome young man would live in princely luxury for one year as the incarnation of the god, after which he would be killed with an obsidian knife and his heart offered to the sun gods. The idea of the centrality of the Sun was still alive and well in more recent times. Consider Louis XIV, king of France from 1643 to 1715, who anointed himself the Sun King and had himself depicted in portraits as the god Apollo.

Even today, when you go to the beach, you're called a "sun worshiper." And note that the wonderful light you feel as you sit there on the beach blanket has left the Sun just eight minutes before. That is how long it takes sunlight to make the 93-million-mile (149-million-kilometer) trip from the Sun. If it were possible to fly a spaceship to the Sun at a speed of 25,000 miles (40,200 kilometers) per hour, it would take about 154 days to make the same trip that sunlight crosses in eight minutes; a jet flying at typical speeds would take seventeen years!

But the solar energy that made that quick trip actually started out in the center of the Sun—its core—millions of years ago. That's how long it takes the energy from the nuclear cauldron that lies at the heart of a star like the Sun to reach the surface.

In one second, the Sun gives off more energy than all people have produced during their stay on Earth. Yet, Earth receives only a tiny fraction—two-billionths—of the Sun's total energy. The rest streams out into space. The real solution to Earth's energy needs will eventually come from devising a way to efficiently capture and harness this nearly unlimited power.

THE SUN: VITAL STATISTICS

Diameter: about 865,000 miles (1,392,000 kilometers), approximately 109 times that of Earth

Distance from the Earth:

  shortest—about 91,400,000 miles (147,100,000 kilometers)

  greatest—about 94,500,000 miles (152,100,000 kilometers)

  mean—about 93 million miles (149 million kilometers)

Age: about 4,600,000,000 years

Rotation period: about 1 month

Revolution period in the Milky Way: about 200 million years at 150 miles (250 kilometers) per second for the Sun to go around the center of the galaxy

Temperature:

  surface—about 10,000°F (5,500°C)

  center—about 27,000,000°F (15,000,000°C)

Mass: 99.8% of the mass of the solar system; about 333,000 times that of Earth

Composition: Hydrogen, about 75%; helium, almost 25%; at least seventy other elements make up the remaining 1 to 2%

How big is the Sun?

The Sun is a bit larger, hotter, brighter, and more massive than the majority of other stars. And, of course, it dwarfs the Earth. About 109 Earths could fit side-by-side across the 864,950-mile (1,392,000-kilometer) diameter of the Sun, and it would require about 333,000 Earths to equal the Sun's mass. *Mass*, to put it simply, is the amount of material something contains. It is similar to weight, except that mass floating in space weighs nothing. That is why astronauts have to undergo weightlessness training.

More than a million Earths could fit inside the Sun.

What sort of star is the Sun?

The Sun is like other stars twinkling in the night sky but it feels hotter and dominates the sky because it is so close to Earth, compared to the next nearest stars; 250,000 times nearer, to be exact. The way in which it generates light and heat is that, every second, 5 million tons of matter are converted into energy by the nuclear reactions in the heart of the Sun, sort of like millions of hydrogen bombs all going off at once. The Sun produces this energy at the rate of 92 billion one-megaton nuclear bombs going off every second.

In ancient times, the Sun was thought to be a ball of fire. Science has updated that view and continues to do so. In recent years, the Sun and other stars were called giant gas balls, but astronomers now say that stars are largely made of *plasma*, a fourth form of matter, which is produced when electrically charged atomic particles (*ions*) come under high heat. On Earth, artificially created plasma is visible when electricity turns the gas in the tube of a neon sign into light, or an arc welder uses electricity to produce high temperatures needed for welding. (This plasma has nothing to do with the biological plasma, which is the clear fluid portion of your blood.)

The energy of the Sun is generated by nuclear-fusion reactions that turn hydrogen, the most abundant element in the universe, into helium at the core. Each time this happens, a minute amount of matter is converted into energy. For each ounce of matter that is annihilated, enough energy is produced to power a 100-watt lightbulb for about 750,000 years.

But that energy isn't visible light yet. It is in the form of high-energy gamma rays, which are invisible to the human eye. Gradually, gamma rays work their way to the Sun's surface, eventually to be changed into visible light. The energy first passes through the *convection zone*, where huge streams of gas carry much of the energy upward. This principle is familiar to earthlings. Hot air rises and cool air sinks—the basic principle that sends hot-air balloons aloft.

The energy then reaches the next layer of the Sun, the visible surface called the *photosphere* ("sphere of light"). It is in this layer that *sunspots*, or visibly dark areas on the Sun, sometimes appear. Sunspots are dark because they are relatively cooler than the rest of the Sun's surface.

The radiation from the Sun travels in a variety of other waves, and each wavelength carries a different amount of energy. Although light-waves are the only visible light, infrared rays can be felt as heat, and ultraviolet rays can tan—or burn—our skin. Some radio waves reach Earth and provide the means of listening to your radio. And we all know what to do with microwaves: Make popcorn quickly! The highest forms of radiation do not pass through the Earth's atmosphere but are now being studied with telescopes and other devices carried aboard satellites.

SOLAR RADIATION, OR THE SUN'S WAVES (FROM THE HIGHEST END OF THE SPECTRUM TO THE LOWEST)

Gamma rays

X-rays

Ultraviolet

Visible

Infrared

Microwaves

Radio waves

Each of these waves has a different frequency—the number of waves that pass by a given point in a given time. That frequency, determined by the number of times per second the electrical charge in the wave vibrates, is measured in hertz (cycles per second), named for German physicist Heinrich Rudolph Hertz (1857–1894).

Above the photosphere is the *chromosphere*, which can be seen as a pink ring around the Sun during an eclipse. This layer is punctuated by *solar flares*—bright, hot jets of gas. Above this is the Sun's atmosphere, the *corona*, a halo of hot gas that surrounds the Sun. Charged particles boiling from the corona stream outward, forming the *solar wind*, an electrified gas, plasma, that moves out through the solar system at about a million miles (1.6 million kilometers) per hour. Activity on the Sun, including sunspots, flares, and prominences, waxes and wanes during the solar cycle, which peaks every eleven years or so and seems to be connected to the solar magnetic field.

Why are there spots on the Sun?

When Chinese astronomers saw dark spots on the Sun's disk one sunset 2,800 years ago, they had no explanation. The great Italian astronomer Galileo said that he first saw these strange markings in 1610, although that got Galileo into trouble with some of the people who argued that they were the first. Galileo did study the sunspots enough to figure out that the Sun was rotating. Observed for centuries, sunspots were once thought to be tornadoes raging across the Sun.

What Galileo, and the Chinese well before him, noticed are blemishes. But these blemishes dwarf a teenager's worst nightmare. These gigantic blotches on the surface of the Sun are often larger than the Earth itself. The largest recorded sunspot, observed in April 1947, was thirty-five times the surface area of the Earth. Galileo could not explain them but made the significant observation that since these spots moved across the face of the Sun, the Sun must be rotating. We now understand that they are "cool spots" on the Sun, where the temperature is 3,500°F cooler than the surrounding area. Sunspots usually occur in groups, up to one hundred at a time, which can last from half a day to several weeks. But, despite the vast range of their size and duration, and their dark appearance, all occur in active zones where the Sun's seething magnetic activity penetrates the outermost layer of hot gas, its photosphere.

These solar blemishes may be very significant to Earth. Studies over long periods show that when there is little sunspot activity, the Earth gets colder. During one seventy-year period, from about 1645 to 1715, when there were few sunspots, Earth went through an unusual cold spell, a brief period that was part of a longer cold spell lasting roughly from 1400 to 1850, which has been called the Little Ice Age. Some scientists believe that glaciers have advanced and retreated in relationship with sunspot activity.

Does the Sun make noise?

Long before Sir Andrew Lloyd Weber's *Phantom of the Opera* made "Music of the Night" a household phrase, there was an idea called "Music of the Spheres." The notion was that the planets and stars actually produce sound as they move through space, an idea that dates to ancient Greece and even earlier. Pythagoras (see part I) taught that the spheres, or planets, made harmonious sounds as they circled the heavens. Plato said that a siren sat on each planet, singing a sweet song that harmonized with all the songs of the other planets. Chaucer, Milton, and Shakespeare all believed in this theory and gave it expression in poetry. The purest scientific approach to this ancient concept was taken by the astronomer Kepler, who laid out his ideas of celestial music in a book called *Harmonice mundi* ("The Harmony of the World") (1619). The first man to understand the motions of planets, Kepler was attempting to figure out the actual notes that planets made as they sped up or slowed down in their orbits. He wrote, "The heavenly motions are nothing but a continual music of several voices, which can be comprehended by the intellect, not the ear." In the book, in which he laid out his Third Law, he said that each planet's speed corresponded to certain notes in the musical scale, and he set down in musical notation his concept of the notes generated by the individual planets as they moved around the Sun. With a flourish, Kepler wrote, "With this symphony of voices man can play through the eternity of time in less than an hour, and can taste in small measure the delight of God, the Supreme Artist. . . . "

But does such music actually exist? Not in the way that Pythagoras thought or Kepler described. Since the early 1960s, astronomers have known that the surface of the Sun pulsates rhythmically. These so-called solar oscillations are caused by low-frequency sound waves that get trapped inside the Sun and reverberate like the vibrations of a ringing bell. Just as geologists can use seismic waves to probe the interior of the Earth, astronomers use solar oscillations to study the Sun's structure.

Since 1996, we have been able to listen to the Sun's "heartbeat" more closely with the Solar and Heliospheric Observatory (SOHO), a kind of stethoscope in space. A joint venture of NASA and the European Space Agency (ESA), SOHO is a 1.85 ton satellite built by the European agency and dispatched by a NASA rocket on December 2, 1995. Commissioned in April 1996 for a nominal operational life of two years, SOHO is expected to function until the end of March 2003, and it now hovers approximately 1 million miles (1.5 million kilometers) from Earth.

In its first five years, SOHO made a huge scientific payoff, including these revelations.

The Sun's surprising heartbeat: Currents of gas far beneath the visible surface speed up and slacken again every sixteen months—a wholly unexpected pulse rate.

Brighter sunbeams: SOHO has seen the Sun brighten, as expected, by 0.1 percent, while the count of sunspots increased 1996–2000. By studying the variations in detail, scientists estimate that high-energy ultraviolet rays from the Sun have become 3 percent stronger over the past three hundred years.

Eruptions coming our way: Most of the explosive outbursts of gas from the Sun, called coronal mass ejections, miss the Earth. Only SOHO can reliably identify those heading in our direction, by linking expanding halos around the Sun to shocks seen in the Earth-facing atmosphere.

Thousands of explosions every day: A reason why the Sun's atmosphere is far hotter than its visible surface is a nonstop succession of small explosions, observed by SOHO. They result from a continual rearrangement of tangled magnetic fields.

The sources of the solar wind: SOHO sees gas leaking from the corners of a magnetic honeycomb of gas bubbles, mainly in polar regions, to supply a fast solar wind. Nearer the Sun's equator, a slow wind escapes from the edges of wedge-shaped features called helmets.

Accelerating the solar wind: Charged atoms feeding the fast wind gain speed very rapidly—evidently driven by strong magnetic waves in the Sun's outer atmosphere.

Elements in the solar wind: SOHO detected phosphorus, chlorine, potassium, titanium, chromium, and nickel for the first time, and previously unseen isotopes of six commoner elements.

Gigantic sunquakes: After a solar flare, SOHO sees waves rushing across the Sun's visible surface, like the ripples seen when a stone falls into a pond.

Huge solar tornadoes: SOHO discovered tornadoes as wide as Africa, with hot gas spiraling outward from the polar regions of the Sun. Typical wind speeds of 31,068 miles (50,000 kilometers) per hour can become ten times faster in gusts.

What's the difference between a solar eclipse and a lunar eclipse?

Eclipse literally means the complete or partial blocking of one celestial body by another. (Astronomers use the word *occultation* for this blocking action.) Although other bodies in the solar system can eclipse each other, from an Earth perspective, eclipses occur when the Sun, Moon, and Earth are all positioned in a straight line. A *solar eclipse* occurs when the Moon passes between the Sun and Earth, temporarily blocking out the Sun's light and darkening all or part of the Sun. A *total eclipse* occurs when the Moon passes between the

Earth and the Sun, completely blotting out the Sun's disk and turning the day into night. In a *partial eclipse*, the Moon covers only a portion of the Sun. A *lunar eclipse* takes place when the Earth passes between the Sun and the Moon, casting a shadow into which the Moon moves.

During a solar eclipse, the Moon's shadow usually moves across the face of the Earth, from west to east, at a speed of about 2,000 miles (3,200 kilometers) per hour. Although the Sun is four hundred times larger than the Moon, the Moon is, coincidentally, four hundred times closer to Earth. That's why the Sun and Moon appear to be the same size—making an eclipse of the Sun possible.

The dark Moon appears on the western edge of the Sun and moves across the Sun. At the moment of total eclipse, a brilliant halo flashes into view around the darkened sun. This halo is the corona or outer atmosphere of the Sun. After a few minutes, the Moon moves off to the east. Although the average eclipse period is about two and a half minutes, the period of total darkness may last as long as seven minutes.

The sight of the Moon completely obscuring the Sun is an impressive one, and has inspired myths and legends throughout history. Eclipse superstition goes back to the earliest days of sky-watching. The ancient Chinese believed that solar eclipses occurred when a dragon in the sky tried to swallow the Sun. In the midst of an eclipse in ancient China, people were called outdoors to make as much noise as possible to frighten away this beast. Since these events became so important in the Chinese world, the Chinese began to chart eclipses, and, over time, discovered a cycle that could be used to predict them. This was not a task taken lightly. In one possibly mythical story, two court astronomers in 2136 B.C. failed to predict an eclipse, a failure that was met with execution.

The Chinese idea of saving the Sun with noise was practiced in Europe as well. From Italy to Scandinavia, people were convinced that the Sun was either going to die or was the victim of an evil spell. They called the noise they made *pandemonium* ("demons everywhere").

In the New Testament, the gospels report an unexpected and long eclipse that coincided with the crucifixion of Christ. During the Middle Ages in Europe, the natural cause of eclipses was understood. However, eclipses were still treated as portents of frightening events, a fact put to use by Christopher Columbus. According to a famous story, the

Italian explorer was stranded in Jamaica in 1504. Columbus carried a book, which included an eclipse schedule, and he intimidated the natives into giving him food by telling them that he would make the sky dark. On February 29, 1504, the lunar eclipse appeared as scheduled, and the terrified natives complied with Columbus's request.

Superstitions surrounding eclipses continued into the twentieth century. In one notable example, people in South America blamed a devastating 1918 outbreak of influenza on a solar eclipse.

A total solar eclipse can be seen only in certain parts of the world. And the most recent total solar eclipse occurred on August 11, 1999, visible in the North Atlantic Ocean, Central Europe, and Central Asia. On June 21, 2001, an eclipse of nearly five minutes was seen from the South Atlantic Ocean, Australia, and Southern Africa.

FUTURE TOTAL SOLAR ECLIPSES

December 4, 2002	Southern Africa, Southern Australia
November 23, 2003	Antarctica
April 8, 2005	South Pacific
March 29, 2006	Africa, Turkey, Azerbaijan, Kazakhstan, Russia
August 1, 2008	Greenland, Siberia, China
July 22, 2009	India, China, Pacific Ocean
July 11, 2010	South Pacific
November 13, 2012	Australia, Pacific Ocean
March 20, 2015	North Atlantic Ocean, Norwegian Sea
March 9, 2016	Indonesia, North Pacific Ocean
August 21, 2017	United States
July 2, 2019	South Pacific Ocean, Chile, Argentina
December 14, 2020	Chile, Argentina

What are the northern lights?

"No pen nor pencil can portray its fickle hues, its radiance, its grandeur," polar explorer William H. Hooper once wrote. He was describing one of nature's most beautiful sights, the *aurora* (the word "aurora" comes from the Latin for "dawn"). Sometimes the aurora looks like a patch of colored-light flashes across the sky. At other times, the lights look like curtains of light caught by a breeze. They come in a palette of colors and, throughout history, they have inspired the poetic in many people. To the Inuit of Alaska and Canada, the *aurora borealis*, or northern lights, was the highest level of heaven, where the dead danced. In Scotland, they have been called "heavenly dancers," "armies fighting in the heavens," and "pools of blood in the firmament." The Romans called the lights "blood rain," and the Chinese "candle dragon." They have been witnessed and marveled at for centuries. In A.D. 37, the Roman emperor Tiberius sent soldiers to put out what was thought to be a fire in the port of Ostia but was really an auroral display. In 1591, a red auroral display over Nuremberg, Germany, provoked a mass panic because people thought that the sky was on fire. During World War II, fighter planes were scrambled to intercept phantom enemy raiders in response to what were auroral displays.

There are actually two different sets of aurorae (the plural of aurora): the aurora borealis, or northern lights, which appear over the North Pole; and the *aurora australis*, southern lights, which appear around the South Pole. Only recently have scientists studying this wonder of nature begun to realize that they are generated by the interaction of charged particles streaming from the Sun in the solar wind and the magnetic field that surrounds the Earth. The Earth's magnetic field deflects most of the deadly particle stream, but the electromagnetic forces draw these particles down into Earth's atmosphere at the polar regions where they collide with atoms and molecules of gas, producing the wondrous light-shows.

Aurorae (pronounced "or-roar-ree") appear in so many colors because each gas in the atmosphere glows a different hue when struck by solar particles. The color also varies according to both the electrical state and concentration of the gas. The most common auroral color, a brilliant green, results when oxygen is struck at low altitudes; The vivid-red aurorae that are seen during major disturbances come when oxygen is struck at higher altitudes. Nitrogen atoms, by contrast, emit blue or purple light, which happens when the atoms are struck by the ultraviolet radiation contained in sunlight.

Why bother with this heavenly light-show? The aurorae are key indicators of solar terrestrial interactions, better known as space weather, which is created by the solar wind. Although it is called "wind," it has nothing to do with the flow of air as it is on Earth. Instead, this wind is a stream of plasma—the charged particles and magnetism—that flows continuously from the Sun. When solar disturbances such as flares erupt, the solar wind gets cranked up to gale force. That sort of high-energy chaos makes for great auroras but it can also play havoc with radio communications, radar systems, power transmission lines, telephone cables, and satellites on which humanity grows more dependent every day. A stunning auroral substorm was blamed for a nine-hour-long power outage that darkened all of Quebec.

Will the Sun always come out?

Yes, Annie. "The sun'll come out tomorrow." And for a few more tomorrows after that. But it will not come out forever.

Every star has a life cycle, and the Sun, born nearly 5 billion years ago, is just shy of its midlife crisis, the starry equivalent of a "thirtysomething." But, eventually, as it nears the end of its life, the Sun will begin to swell, gradually reaching up to one hundred times its current radius or reaching out as far as the current orbit of Mars. In this phase, stars are known as red giants. As its outer layers dissipate into space, the Sun will shrink to its compact dense core and become a white dwarf about the size of Earth. (Star terms such as "red giant" and "white dwarf" are discussed in greater depth in part III.)

What happens to the planets in this scenario? As the Sun expands to the size of a red giant, it will actually lose mass, and its gravitational pull on the planets will weaken, and they will start to wander away. (Remember: the Greek word for planet is "wanderer.") Only Mercury, the planet closest to the Sun, will be Sun-grilled. Earth will escape the Sun's pull along with the other planets. But only after everything on Earth has been fairly well scorched. Of course, you needn't worry about this scenario. After all, the Sun has somewhere in the neighborhood of another 6 billion years to live before it begins to go through these agonizing death throes.

MERCURY

If you went to Mercury

Small, bleak, dark, and gray. No, it isn't an overpriced Manhattan studio apartment. Barren Mercury is the planet closest to the Sun, the second smallest planet; only distant Pluto is smaller. Because it orbits the Sun so quickly—in just eighty-eight days—Mercury was named for the Roman messenger of the gods and the god of commerce. (In Greek mythology, he was called Hermes.) Although its orbit is rapid, its rotation is not. Mercury takes almost fifty-nine Earth days to complete one rotation.

Scorched by solar radiation and with no significant atmosphere, the surface of Mercury can reach a searing 800°F during the day and plunge to –290°F at night. These conditions have made any sort of probes landing on Mercury highly difficult. But, in the 1970s, the spacecraft *Mariner 10* flew by Mercury and sent back pictures of a rocky planet, which is similar in many respects to the Earth's Moon. The small rocky planet is scarred by impact craters and is close to the Moon in size, a little under 1.5 times larger. With no atmosphere, Mercury's surface has remained unchanged for billions of years. Its major features include *scarps*, or huge, twisting ridges that snake across the surface, sometimes for more than 100 miles—and sometimes rise to nearly 10,000 feet. These may date from a time when the planet "shrank," perhaps as its core cooled. Mercury's largest known feature, the Caloris Basin, a crater roughly the size of Texas, is thought to be the result of a rock slamming into the planet. Mercury's craters, which come in all sizes, were created when space rubble crashed into Mercury early in the history of the solar system, some 4 billion years ago. The remnants of these impacts are white, powdery streaks that spread out from the craters, many of which have been named in honor of Earth's greatest artists, writers, and composers (Beethoven, Raphael, Shelley, Keats, Vivaldi, etc.).

MERCURY: VITAL STATISTICS

Diameter: 3,031 miles (4,880 kilometers)

Mean distance from the Sun: 36 million miles (58 million kilometers)

Distance from the Earth:

  shortest — 57,000,000 miles (91,700,000 kilometers)

  greatest — 136,000,000 miles (218,900,000 kilometers)

Length of year: 87.97 Earth days

Rotation period: 58.65 Earth days

Surface temperature: 801°F to –279°F (427°C to –127°C)

Atmosphere: oxygen, sodium, hydrogen, helium

Probable composition: large iron core, mantle of silicate materials

Known satellites: none known

Why is it difficult to see Mercury?

Mercury is only visible for about one hour before sunrise and one hour after sunset. Because of its proximity to the Sun, there is intense glare that makes it difficult to see Mercury, which can never be seen in a fully dark sky.

VOICES OF THE UNIVERSE:
Native American chant celebrating Venus

The morning star looks like a man who is completely covered in red paint; that is the color of life. He wears leg warmers and a cloak. On his head lies a soft and downy eagle's feather, painted red. This feather represents the soft light cloud that floats high in the sky. . . . Morning star give us strength and renewal. . . . The day is on its heels.

VENUS

If you went to Venus

Among the heavenly bodies, perhaps only the Sun and Moon have inspired more myths than Venus. The second planet from the Sun is named for the Roman goddess of love and sensual pleasure. Most of its features have been given female names. Its two main landmasses, or continents, Ishtar Terra and Aphrodite Terra, have been named for the Babylonian and Greek goddesses of love.

But others saw Venus as a malevolent planet. In ancient Mexico, people locked their doors and windows before dawn to protect themselves from this planet which, they imagined, carried dreadful diseases. (*Venereal* disease comes from the same Latin word as Venus.) For the Maya and Aztecs, Venus was a symbol of death and rebirth.

With an orbit that brings it within 26 million miles (41 million kilometers) of Earth, Venus is closer to us than any other planet. Both planets were formed at about the same time from similar material, and once had similar atmospheres. Venus is also similar to Earth in size and mass, a fact which once inspired images of a lush, habitable sister planet. But a generation of probes have given us a good sense of Venus (99 percent of its surface was mapped by the *Magellan* spacecraft), and we know it to be lifeless and dry, with no trace of water on its surface. Any water Venus may have once had was evaporated long ago by the intense surface heat. Searing temperatures, crushing pressures, and a suffocating, poisonous atmosphere combine to make Venus a very different world from ours.

If you could go, you would see a landscape of massive volcanoes, surrounded by extensive lava plains crossed by lava-flow channels thousands of miles long. The few impact craters are large because only the most massive meteorites have been able to penetrate the atmosphere.

VENUS: VITAL STATISTICS

Diameter: 7,521 miles (12,104 kilometers)

Mean distance from the Sun: 67,230,000 miles (108,200,000 kilometers)

Distance from the Earth:

shortest — 25,700,000 miles (41,400,000 kilometers)

greatest — 160,000,000 miles (257,000,000 kilometers)

Length of year: 225 Earth days

Rotation period: 243 Earth days

Temperature: 864°F (462°C).

Atmosphere: Carbon dioxide, nitrogen, water vapor, argon, carbon monoxide, neon, sulfur dioxide

Composition: surface consists mostly of silicate rock and may have an iron/nickel core, similar to that of Earth

Satellites: none known

Since Venus takes 243 days to turn once on its axis and 25 days to go around the Sun, it is the only planet that takes longer to turn than to orbit. On Venus, a day is longer than a year.

Relative to the other planets in the solar system, Venus spins backward on its axis. Since Venus spins to the west, the Sun rises in the west and sets in the east. The reason for this backward, or retrograde, motion is unknown. One widely held theory is that early in its history, when all the planets were forming, Venus may have been hit by a massive object that reversed its spin.

Why is Venus so hot?

Scientists think the original atmosphere of both Venus and Earth were created from gases released by volcanoes, when both planets were very young and volcanic activity was much more intense. But because Venus is so close to the Sun, the "greenhouse effect," in which heat is trapped within its atmosphere, results in the temperature rising so high that all the surface water evaporated. With all the water now in the atmosphere, the intense ultraviolet radiation from the Sun split the water molecules into hydrogen and oxygen. The hydrogen escaped into space and the oxygen combined with other chemicals in the atmosphere. In contrast, Earth cooled down, oceans formed, and life began to develop. Earth became a living planet while Venus, despite its connection to the goddess of love and fertility, remained barren.

If Venus is a planet, why is it called the "Evening Star"?

This is simply one of those cases where common parlance takes precedence over scientific accuracy. After the Moon, Venus is the most brilliant object in the night sky, and early observers thought it was a bright star. Venus appears so bright because it is near to Earth, and its cloud-topped surface reflects 75 percent of the light that falls on it. Venus is also often called the "Morning Star," because it is most visible just before sunrise as well as sunset.

What is the transit of Venus?

If you answer "An erotic novel by Anais Nin," your head may not be in the stars. (Nin's erotically charged book is called *Delta of Venus*.)

Both Mercury and Venus can sometimes be seen as dots passing in front of the Sun. Such an event is called a transit. Kepler was the first to recognize that transits must exist, and to calculate when they would occur. With Kepler's calculations in hand, Pierre Gassendi (1592–1655) was able to observe the transit of Mercury in 1631, a year after Kepler's death. Jeremiah Horrocks improved on Kepler's calculations and became the first to observe a transit of Venus on November 24, 1639. It was Horrocks who realized that if a transit of Venus were observed simultaneously from several places on the Earth, the information could be used to calculate both the distance to Venus and the distance from Earth to the Sun.

In 1761, astronomers were dispatched to India, St. Helena, and other viewing spots. War and cloudy skies prevented some of the necessary observations from being made, but a few observations were made, and it was during this transit that the atmosphere of Venus was discovered.

The most famous transit of Venus occurred eight years later, when the English explorer Captain James Cook undertook the first leg of his great voyages of discovery to observe this transit from Tahiti. Leaving England on the *Endeavour* on August 26, 1768, Cook made the necessary observations in Tahiti and returned to England on July 17, 1771. Cook's was not the only trip of discovery based on the 1769 transit. The Russians journeyed overland to Siberia to observe the event, and other observers were sent to various points around the globe.

In the mid-nineteenth century, Johann Franz Encke (1791–1865) used this transit data to calculate the Earth's distance to the Sun as 153,000,000 kilometers (95,300,000 miles), the best calculation to that time.

VOICES OF THE UNIVERSE:
Genesis 1:1–10

> *In the beginning God created the heaven and the earth. And the earth was without form, and void; and darkness was upon the face of the deep. And the Spirit of God moved upon the face of the waters.*
>
> *And God said, Let there be light: and there was light.*
>
> *And God saw the light, that it was good: and God divided the light from the darkness.*
>
> *And God called the Light Day, and the darkness he called Night. And the evening and the morning were the first day.*
>
> *And God said, Let there be a firmament in the midst of the waters, and let it divide the waters from the waters.*
>
> *And God made the firmament and divided the waters which were under the firmament from the waters which were above the firmament: and it was so.*
>
> *And God called the firmament heaven. And the evening and the morning were the second day.*
>
> *And God said, Let the waters under the heaven be gathered together unto one place and let the dry land appear: and it was so.*
>
> *And God called the dry land Earth; and the gathering together of the waters called he Seas; And God saw that it was good.*

EARTH

EARTH: VITAL STATISTICS

Diameter (at the equator): 7,926.41 miles (12,756.32 kilometers)

Equatorial circumference (distance around Earth along the equator): 24,901.55 miles (40,075.16 kilometers)

Mean distance from the Sun: 93 million miles (150 million kilometers)

Length of year: 365 days, 6 hours, 9 minutes, 9.54 seconds

Length of day: 23 hours, 56 minutes, 4.09 seconds

Average surface temperature: 57°F (14°C)

Composition: iron, nickel, silicon, aluminum

Atmosphere: nitrogen, oxygen, argon, with small amounts of other gases

Satellites: 1 (and many artificial ones!)

- Moon

Why is there life on Earth?

Remember Goldilocks and Baby Bear's porridge? Not too hot. Not too cool. Just right. Earth is, essentially, the Baby Bear's porridge of the solar system—and, as far as we know, the entire universe. Its distance from the Sun is close enough to be warmed without becoming too cold; but not too close as to get too hot.

Originating as molten magma 4.5 billion years ago, Earth eventually cooled down and solidified. Fissures cracked in the hardening crust, and lava and trapped gases bubbled to the surface, forming Earth's primitive atmosphere. Water vapor eventually condensed, rose, cooled, and fell as rain. The Genesis account, in which the land followed the seas, is actually a neatly poetic version of this scenario. Except that it rained for a long time. As time passed, the crust got waterlogged, and the falling rain collected in vast basins and became the first seas. New evidence suggests that Earth's temperature may have cooled sufficiently to retain the oceans 4.4 billion years ago. Recent studies of a tiny grain of zircon show that about 100 million years after Earth was formed from the bits and pieces left over after the Sun's birth, temperatures on the infant planet had cooled to near or below the boiling point of water, significant because liquid water is considered a precondition for life.

Sunlight fell on the water and land. And it was good. Really good. Sun, water, and electricity in the form of lightning were some of the ingredients of the stew in which life slowly emerged after a billion years or so. Based on analysis of micro fossils, scientists believe that about 3.5 billion years ago, primitive bacteria and, later, blue green algae, plantlike organisms, began to grow in the seas. Over millions of years, some of these developed into plants that could use sunlight to make their own food—what is called *photosynthesis*. The plants took in water and carbon dioxide and converted them into sugar. The waste products of this process were water and oxygen. They were so good at doing this, that these plants eventually changed the atmosphere. Some of that oxygen accumulated, forming the air we breathe today.

How did we get those early complex molecules floating around—nitrogen, oxygen, carbon dioxide gas, ammonia, and methane—to grow into bacteria and plants and, eventually, you and me? There are only theories. But it must have happened something like this. The turbulent atmosphere of early Earth, along with the violent rumblings of volcanoes and earthquakes, as well as continuous bombardment from space, provided a giant Cuisinart of life. In other words, Mr. Bond, the early Earth was shaken, not stirred. A major new area of investigation returns the search for answers to the seas of Genesis. Deep under the oceans, volcanic vents spew forth heat and an intriguing mixture of chemicals. Could these volcanic vents be the wellspring of life?

As hundreds of millions more years passed, water plants crept up onto the dry land. So did the fishlike organisms that had begun to evolve in the seas about 350 million years ago. Fast-forward to 225 million years ago, and some of those fishes who had managed to crawl out of the water adapted to their new surroundings and gradually became reptiles—the greatest of which were the dinosaurs. Then something happened about 65 million years ago, and most of them disappeared in one of the mass extinctions that have punctuated Earth's long history. Was it something they ate? Not enough exercise? Too much cholesterol? Or an asteroid?

The little guys who survived that extinction eventually grew up in the age of mammals—the age in which we humans still live. But we are newcomers; our ancestors have only been around for a brief few million years or so. While we have a growing fossil record to date the lifestyles of the hairy biped, it was only about fifty thousand years ago—a blink of the cosmic eye—that evidence of human expression appeared. And one of the first things those prehistoric ancestors did was carve pictures in stones. On some of them you can make out pictures of stars—a clue that even the earliest humans were fascinated by the heavens.

So, are we and our planet unique? So far, yes. We haven't found anything else in the universe quite like the "third rock from the Sun" and its ideal conditions for spawning life—as we know it. But as archeologists like to say, "Absence of evidence is not evidence of absence." Or, just because we haven't found another potential Earth yet—since we have literally just begun to scratch the minutest surface of the rest of the universe—doesn't mean there aren't more Earths out there.

VOICES OF THE UNIVERSE:
CHARLES ROBERT DARWIN (1809–1882)

> We may well affirm that every part of the world is habitable!
> Whether lakes of brine, or those subterranean ones hidden
> beneath volcanic mountains—warm mineral springs—the
> wide expanse and depths of the ocean—the upper regions of
> the atmosphere—and even the surface of perpetual snow—
> all support organic things.

Does Earth wobble?

Feeling a little woozy? Like the whole room is spinning around? If the answer is "yes," don't blame the Earth. Yes, it does not spin perfectly but wobbles a bit, like a top that is starting to slow down. The Earth has a very slight wobble at the poles, called the *precession of the equinoxes*. This explains why the North Star commonly seen today, Polaris, the star that remains fixed above the North Pole, is not the same North Star the Egyptians looked for in 3000 B.C., which they called Thuban. Not to worry—this takes a long time, like 27,000 years. Eventually, the North Star, Polaris, won't be the North Star anymore. Its place will be taken by another star, called Vega. Don't worry about changing your star charts just yet, though. This doesn't happen for another 12,000 years.

VOICES OF THE UNIVERSE:
NEIL ARMSTRONG, July 20, 1969

On reaching the Moon: "Houston, Tranquility Base here. The Eagle has landed."

On first stepping on the Moon: "That's one small step for man, one giant leap for mankind."

THE MOON

If you went to the Moon

The Moon, Earth's closest companion, travels with the Earth in space around the Sun, the most familiar sight in the night sky. A gray desert, the Moon is dotted with craters from ancient asteroid collisions. Unlike the other planets in the solar system, when it comes to our Moon, we can actually answer the question "What if you went?" So far, the Moon is the only body in space on which humans have landed. It has been thirty years since we sent astronauts to the Moon, but a recent discovery of water there has stimulated plans for an eventual return.

The Moon's chief features include:

Craters: The surface of the Moon is pitted with thousands of impact craters, most of them formed long ago by meteorites crashing into the Moon. Some are enormous, including the Copernicus crater (57 miles, or 91 kilometers across) or Tycho (54 miles, or 87 kilometers across). Craters at the poles never receive sunlight and, in 1998, were found to contain traces of frozen water. Although most craters were formed billions of years ago, there is one relatively recent account of a crater being formed. In 1178, a group of five British monks were observing the moon. According to a written account from that period, "suddenly the upper horn split in two. From the midpoint of this diversion a flaming torch sprang . . . spewing out, over a considerable distance, fire, hot coals, sparks." The monks had most likely seen a meteor crashing into the lunar surface. Now named after Giordano Bruno, the crater left by the impact is relatively small—about 13.5 miles in diameter—and was probably caused by a fairly small meteoroid.

Maria (or seas, plural of the Latin *mare* for "sea"): The darker areas of the Moon's surface. They are depressions created by the impact of giant meteorites, which were filled with dark lava that broke flow through the floors of these basins and spread over the surface.

Highlands: The bright areas on the Moon are craggy highlands, whose light rocks reflect the sunlight. This contrasts with the dark, sunlight-absorbing rock in the lava filled maria.

Rays: Some craters are surrounded by bright lines that radiate outward and are formed from the powdery debris thrown out of the crater at the time of impact. These rays consist of crushed rock.

Moon dust: The Moon is covered in very fine dust called *regolith*. It is made from rock pulverized by eons of meteor impacts. Some pieces of rock are shaped like droplets, where the rock has been melted and splashed across the surface.

LUNAR DIET

1ST WEEK   2ND WEEK   3RD WEEK

John O'Brien

How long is the Moon's month: 27 days or 29 days?

Astronomy books and lunar calendars sometimes mention two different figures for the length of the lunar month. This seeming confusion comes from the fact that scientists measure the Moon's revolution around the Earth in two different ways: *synodic months* and *sidereal months*. Most people typically think about the phases of the Moon in terms of the synodic month, which equals about 29.5 days. That is the length of time from one new Moon to the next. It is the time that the Moon takes to revolve around the Earth in relation to the Sun. If the Moon started on its orbit from a spot exactly between the Earth and Sun, it would return to almost the same place in 29.5 days.

But a sidereal month—about 27.33 days—is the time the Moon takes to make one trip around the Earth in relation to the stars. If you marked the beginning of the Moon's revolution in line with a particular star, it would return to the same position about 27.33 days later.

A synodic month is longer than a sidereal month because the Earth travels around the Sun while the Moon travels around the Earth. By the time the Moon has made one revolution around the Earth, the Earth has revolved one-thirteenth of the way around the Sun. That means the Moon has to travel slightly farther to be in the same position in relation to the Sun.

As it revolves around the Earth, the Moon passes through phases of reflected sunlight. When the Moon is between the Sun and the Earth, we can't see it, because the sunlight is striking its far side. That is the new moon. As it moves around the Earth, sunlight begins to strike the side visible to the Earth in a growing proportion. When the Moon is at a right angle to the Earth (the first quarter), we see a half-moon. When the Earth lies between the Sun and Moon, the Moon's face is fully illuminated by the Sun, and we see a full moon. Then it moves to a right angle on the other side as it wanes, or begins to disappear. In the last quarter, we see a half-moon, then a crescent, until the Moon comes full circle, back to new moon status. These phases occur most of the time, as the Earth and the Moon are not aligned in a perfectly straight line. When that does happen, the Earth blocks the sunlight from reaching the Moon, and the result is a total lunar eclipse.

VOICES OF THE UNIVERSE:
Enouma Elish, Babylonian Creation epic, circa 1700 B.C.

> At the beginning, the gods Anou, Enlil, and Ea divided everything up between the two gods who were the keeper of the sky and the earth . . . Sin and Shamash were given two equal portions, day and night.

What's up with the Man in the Moon?

Along with the Sun, the Moon has been the most inspirational heavenly body, its proximity, regularity, and beauty firing the human imagination in myth, religion, poetry, and art. It has also inspired some silliness, such as the idea of a "Man in the Moon," and being made of green cheese.

The first people to write about the Moon were the Babylonians, whose Creation stories known as *Enouma Elish*, composed almost four thousand years ago, said that the Sun and Moon were born together. Some Mesoamerican civilizations had the same idea. To the Babylonians, the Moon, which goes through its regular phases of appearing, growing larger, then disappearing again (waxing and waning) was a perfect symbol of life. The Moon's cycle, so similar to the cycles of human female fertility, became closely connected to fertility rites.

Because the Moon and the Earth are locked in an eternal dance in which the Moon's rotation is similar to that of Earth's, we always see the same side of the Moon. Throughout history, people saw a human face in that Moon, a "Man in the Moon." Some saw his "features" in the maria. The *Mare Nubium* was his mouth, and the eyes were formed by the *Mare Serenitatis* and the *Mare Imbrium.* One popular concept held that the Man in the Moon was an actual man who had been transported to the Moon and was nailed there as punishment and to provide a threatening example.

Where did the Moon come from?

We don't really know, but there are two best guesses. One of these theories holds that the Earth captured a small planet in its gravitational pull during the earliest period of the solar system's formation. However, most scientists believe in what is usually called the *giant-impact theory.* Based on the study of rocks returned by the Apollo Moon missions, the Moon is actually formed from a piece of Earth, since they are approximately the same age. In this widely accepted view, the young Earth was being pummeled for 500 million years by intense meteor showers—the leftover bits and pieces of the solar system's formative basic components. About 4.45 billion years ago, an object the size of Mars may have slammed into the recently formed Earth, blasting out fragments of the young planet in the form of vaporized hot rock. Thrown into space, these fragments gradually came together, were caught in an orbit around Earth, and eventually combined to form the Moon.

Is a "blue moon" really blue?

"Blue moon / You saw me standing alone."

Rogers and Hart were certainly thinking poetically rather than astronomically when they penned those famous lyrics. A blue moon has nothing to do with color. But more surprising is the fact that the usual explanation of a blue moon is also less than accurate. For a long time, most people believed that a "blue moon" occurs whenever there is a full moon twice in one month. However, in 1999, *Sky & Telescope* magazine reported that the mistaken assumption was a result of an error in an article published by the magazine in 1946. The precise definition, based on an old *Farmer's Almanac*, is that when one of the four seasons contains four full moons, the third is called *the blue moon*.

The Moon's cycle is not exactly the same each month, varying from 29.2 to 29.9 days, averaging 29.53 days. The expression "once in a blue moon," meaning something that occurs only rarely, is a bit off the mark. Blue moons are quite predictable and generally occur every three years. The year 1999 was unusual for blue moons, because there were two of them in the same year within the space of three months: two full moons in January, no full moon in February, and two in March. A sequence like that had not happened in more than eighty years. There will not be two blue moons in the same calendar year again until 2018.

As for the color question, there are occasions when the Moon has a bluish tinge. But that coloration is more likely due to unusual atmospheric conditions, such as high amounts of dust in the air, perhaps following a volcanic eruption or extensive forest fires. Soot particles, deposited in the Earth's atmosphere, selectively absorb parts of the red-light spectrum, accounting for the color change. A notable 1950 blue moon observed around the world may have been produced by heavy forest fires in Canada.

So if a blue moon isn't really blue, why do we call it that? A proverb first recorded in 1528 said:

Yf they saye the mone is belue
We must believe that it is true.

The proverb's context meant "don't put stock in an absurdity"—it was ridiculous to think of the Moon as blue because it never happened. As time went by, however, the meaning of a "blue moon" evolved from something that never happened, to an extremely rare event.

So how many blue moons can we count on in the coming years? The following chart lists them:

Year	Date
2001	November 30
2004	July 31
2007	June 30
2009	December 31
2012	August 31
2015	July 31
2018	January 31
2018	March 31
2020	October 31
2023	August 31
2026	May 31
2028	December 31

Source: Sten Odenwald, *The Astronomy Cafe*

VOICES OF THE UNIVERSE:
WILLIAM SHAKESPEARE, from
A Midsummer Night's Dream

The lunatic, the lover and the poet are of imagination all compact.

Who were the lunatics?

In 1764, the Lunar Society was founded in Birmingham, England. Although it has nothing to do with studying the Moon, the Lunar Society occupies a special place in the history of science and the industrial revolution of England. Its membership included James Watt, the inventor of many improvements on the steam engine; Joseph Priestley, one of the great chemists of the period; Josiah Wedgwood, the potter who not only made the plates for which he is legendary, but also various advances in ceramics that had important scientific applications; William Withering, who introduced digitalis for the treatment of heart disease; William Murdock, who developed coal gas for lighting; and America's Benjamin Franklin, a corresponding member.

An eighteenth-century version of Mensa, the society of geniuses, the group met to discuss scientific advances and exchange information. As they held monthly meetings on the Monday night nearest the full moon, so that members could easily see their way home, they hit upon their name, but the members were called lunatics. The word *lunatic*, literally meaning a "moonstruck person," was used in English as early as 1290, but its use goes back to Roman times, when it was popularly believed that the mind is affected by the Moon.

Legends of increased business in the emergency rooms and myths about werewolves on full moons to the contrary, there is no scientific evidence that the full moon has any noticeable effect on human behavior.

Is the Moon Earth's only satellite?

The Moon is, by definition, a satellite. Any body that orbits a planet is a satellite, and there are many more moons orbiting other planets in the solar system. At latest count, there were more than sixty, and the number keeps rising with more sophisticated tracking devices and better telescopes. Orbiting telescopes, like the Hubble Space Telescope, are also satellites, though obviously man-made. Indeed, every time we launch an object that stays in orbit around the Earth, it is considered a satellite.

Man has put up quite a few satellites since the dawn of the Space Age. The United States Space Command reported that in September 2000 there were 2,698 man-made satellites aloft, orbiting the Earth. Only about a third of them still function; the others are "dead" craft that have run out of fuel. As the old saying has it, "What goes up must come down." And the orbits of these "dead ducks" in space eventually begin to deteriorate. Pulled back by Earth's gravity, they fall back toward the Earth and burn upon reentering the atmosphere. Usually.

Two satellites, a Soviet space station and the American Skylab, fell back to Earth but did not disintegrate. The Soviet spacecraft crashed into the Andes, and the American Skylab came down in a remote section of the Australian outback. In April 2001, the Soviets brought down the aging and infirm Mir space station, after fifteen years in space, for a watery burial in the Pacific Ocean.

THE MOON: VITAL STATISTICS

Diameter: about 2,160 miles (3,476 kilometers)

Circumference: about 6,790 miles (10,927 kilometers)

Distance from the Earth:

   shortest—221,456 miles (356,399 kilometers)

   greatest—252,711 miles (406,699 kilometers)

   mean—238,857 miles (384,403 kilometers)

Age: About 4.6 billion years.

Revolution period around the Earth: 27 days, 7 hours, 43 minutes

Average speed around the Earth: 2,300 miles (3,700 kilometers) per hour

Length of day and night: about 15 Earth days each

Temperature at equator: Sun at zenith over maria—260°F (127°C); Lunar night on maria——280°F (–173° C)

Atmosphere: Little or none

VOICES OF THE UNIVERSE:
H.G. WELLS, *War of the Worlds*, 1897

No one would have believed in the last years of the nineteenth century that this world was being watched keenly and closely by intelligences greater than man's and yet as mortal as his own; that as men busied themselves about their various concerns, they were scrutinised and studied, perhaps almost as narrowly as a man with a microscope might scrutinise the transient creatures that swarm and multiply in a drop of water. With infinite complacency, men went to and fro over this globe about their little affairs, serene in their assurance of their empire over matter. . . . No one gave a thought to the older worlds of space as sources of human danger, or thought of them only to dismiss the idea of life upon them as impossible or improbable. . . . At most, terrestrial men fancied there might be other men upon Mars, perhaps inferior to themselves and ready to welcome a missionary enterprise. Yet across the gulf, minds that are to our minds as ours are to those of the beasts that perish, intellects vast and cool and unsympathetic, regarded this earth with envious eyes, and slowly and surely drew their plans against us.

On Halloween of 1938, much of the eastern United States was thrown into a thorough panic when a radio version of *The War of the Worlds*, produced and directed by Orson Welles, was broadcast in a live performance that had the immediacy of a news broadcast. Millions of Americans, already wary because of the fears of an approaching war in Europe, were frightened into believing that Martians were attacking.

MARS

If you went to Mars

The fourth planet from the Sun, one of Earth's nearest neighbors, Mars is a small, rocky planet that has become more fixed in the human imagination than most other planets. Science-fiction writers, such as H. G. Wells, and Hollywood have always found Mars a fertile source for tales of dread. The threatening image of Mars, which was named for the Roman god of war, is derived from its color in the sky. It is noticeably red, so the connection to blood and destruction was made early in human history. Colder than the Earth and just over half the size of Earth, Mars is otherwise remarkably similar to our planet.

Tilted on its axis, like the Earth, it has days and seasons, a thin atmosphere and—as we have begun to learn—quite possibly, significant reserves of water buried as subterranean ice. Mars has a relatively warm summer, when temperatures in the southern hemisphere can reach 68°F, but a long, cold winter in which temperatures plunge to –284°F.

Billions of years ago, Mars was covered with massive volcanoes—just like the Earth—and possibly had some surface water, which gathered in flash floods, carving water channels into the terrain. After the Moon, Mars has been explored to a greater extent than any other part of the solar system. But it has not been without difficulty. We often think of Mars as a stationary object in space, but, remember, it is a moving target—just as we are. Sending spacecraft across 200 million miles of space is not like tossing apples into a bushel basket. The human track record on getting to Mars is proof of that. Since 1960, eighteen Mars probes—four American and fourteen Russian—had failed to reach Mars. Five Soviet probes in the early sixties never got out of Earth's orbit. America's *Mariner 3* went into a useless orbit around the Sun. During the 1970s, five Soviet Mars probes failed for a variety of reasons. The Soviet Union launched two craft to Mars in 1988. One became lost in space. Communications with the other broke down shortly before its scheduled landing in 1989. The United States launched the Mars Observer probe in September 1992. On August 21, 1993, three days before the probe was to go into orbit around Mars, mission controllers on Earth lost contact with it.

But, if at first you don't succeed . . . On December 4, 1996, the United States launched the *Pathfinder* probe, which landed on Mars on July 4, 1997, to become one of the most spectacularly successful space missions since the days of the Apollo Moon missions. A high point in the drama occurred two days later, on July 6, following a short delay, when a small, six-wheeled vehicle called *Sojourner* rolled down a ramp from *Pathfinder* to the Martian surface. The *Sojourner* was a remote-controlled rover that looked like it had been constructed out of some really brilliant child's Erector set. But the world was soon seeing pictures of the ancient red planet, ten minutes after they were sent back to Earth. What millions of people saw and marveled at was a landscape that looked a bit like the dry, rugged, brownish desert of the

American southwest. Only there were no fast-food joints selling burritos or gift shops selling cheap souvenirs, as in "My rover went to Mars and all I got was this crummy T-shirt."

The search for water—and life—on Mars has been one of the driving forces of the recent American space program. But *Pathfinder/Sojourner*'s great success was followed by a string of embarrassing disasters, as NASA's two most recent efforts to explore Mars led to spacecraft being lost before reaching Mars. In one of NASA's most red-faced moments, the Mars Climate Orbiter strayed off-course in 1999. Apparently, a mission controller failed to convert some measurements from linear to metric units. Yes, there is a difference between inches and centimeters. Or, as the old carpenter's rule goes, "Measure twice, cut once." Then, later in 1999, as millions of people anticipated the touchdown of the Mars Polar Lander that could be tracked in "real time" over the Internet, the much-hyped moment fizzled. The lander never sent back a signal. Whether it crashed before landing, or landed, and its communications equipment failed, remains

a mystery that will probably have to wait for the next planned Mars fly-by or landing. For the once-proud space agency, the Mars fiascos became a scientific and public-relations catastrophe.

The major features of Mars include:

North Pole: Made of frozen carbon dioxide ("dry ice") and water ice. The southern ice cap shrinks away to almost nothing in the summer.

Olympus Mons: The largest volcano in the solar system. Most likely extinct, it rises 17 miles above the plains and is 375 miles across.

Valles Marineris Canyon: A giant canyon across one side of Mars, 3,125 miles long and so big that the Rocky Mountains would fit comfortably inside.

Argyre Planitia: One of many basins created billions of years ago by asteroid impacts. The crater Galle is about 125 miles across.

Mars also has two tiny moons—Phobos and Deimos—which race around the planet in about eight hours and thirty hours respectively. In Greek mythology, Phobos and Deimos were the sons of Ares (Mars) and Aphrodite (Venus) and attended their father. Phobos was the god of panic on the battlefield, and his brother Deimos was the god of fear.

Is Mars really red?

Reddish, rust-brown areas cover about two-thirds of the Martian surface. They are dry, desertlike regions covered by dust, sand, and rocks. Much of the surface material seems to contain a brick-colored mineral called *limonite*, which is found in Earth's deserts. Continuous winds blow across the Martian surface at 125 miles per hour, whipping up storms of fine orange-brown dust. It is this dust that earned Mars the name "the Red Planet." The color is not fire-engine- or blood-red, but rusty red, as the color comes from the high proportion of iron in the planet's rocks—twice as much as on Earth.

Who dug those Martian canals?

The greatest long-standing myth about Mars regards the planet's so-called "canals." The idea of actual canals that had been dug by creatures on Mars has been stuck in the human imagination for a long time, but they are actually the result of a very earthly goof: a simple mistranslation. Observing Mars in 1877, Italian astronomer Giovanni Schiaparelli (1835–1910) saw what he described in Italian as *canali*, meaning "channels," through his telescope. In a later English version of his work, the word was mistranslated as "canals," establishing the notion that they must be artificial waterways created by some advanced society of Martians. In fact, they are natural features, such as craters and canyons.

But Percival Lowell, a member of the wealthy and distinguished Lowell family of Massachusetts, was intrigued by the idea. Using some of his family's wealth, Lowell established a private observatory in Flagstaff, Arizona, and commenced to study Mars, publishing *Mars and Its Canals* in 1906. Noting that the canals seemed to change with the seasons, Lowell believed that the advanced beings on Mars had devised a way to transport water from the Martian ice caps.

Lowell's "canals," or Schiaparelli's *canali* were later shown to be an optical illusion.

Mars: Desert or wetland?

No canals. But that doesn't mean "no water." In recent years, there has been more investigation and speculation about water on Mars than almost any other question confronting astronomers. Late in the year 2000, NASA announced that photographs taken by the Mars Global Surveyor offered strong evidence that the planet was once a watery place. Layers of sediment within basins may possibly be the remains of great lakes that once filled the ancient craters on Mars. At the Martian north pole there is water locked in ice; at the South Pole, frozen carbon dioxide (dry ice).

If this is true, it leaves an intriguing puzzle: Where did the water come from? One theory is that it was underground, in the planet's hot interior, and bubbled out. Studies of Martian rocks—meteorites found on Earth that are believed to have originated on Mars—had led to the conclusion that the magma (rock formed by cooling lava from volcanic eruptions) was dry, but then a recent study reported in the journal *Nature* in 2000 concluded that there may have indeed been water in the magma. The new conclusion came about from studying a 175-million-year-old meteorite named Shergotty.

The revised theory is based on what we know about volcanoes. In a volcano—and Mars has many towering volcanoes, including the largest one in the solar system—hot molten lava is generated deep down. As it rises and meets cooler rocks, which contain hydrogen, the magma melts these rocks, releasing the hydrogen to form water.

Mars: Dead or alive?

In 1984, scientists discovered a potato-sized meteorite, a piece of basalt designated ALH84001, found in the Allen Hills (thus the ALH designation) ice field in Antarctica. Carbon dating showed the rock to be 4.5 billion years old. Presumably, this potato-sized chunk of rock was dislodged from the surface of Mars when the planet was hit by an asteroid or comet and the debris caused by the impact was blasted into space. After floating in space for thousands of years, the rock was captured by Earth's gravity, landing in Antarctica between 11,000 and 13,000 years ago. Scientists believe the rock originated on Mars, because it is chemically similar to soil analyzed by space probes that landed on Mars in 1976.

After sitting quietly in a laboratory for most of a decade, the rock created international front-page news long after its discovery. At a press conference on August 6, 1996, a NASA team led by David S. McKay reported that they had found evidence in the rock of microscopic fossils resembling ancient Earth bacteria—suggesting that there was life on Mars more than 3.6 billion years ago.

But we are not talking about Marvin the Martian, invaders from Mars, or little green men with ray guns here. Using a scanning electron microscope, the team saw what appeared to be elongated bodies on the rock's surface. In what was an astonishing discovery, they seemed to resemble nanobacteria, Earth's smallest living organisms. These tiny tubular and egg-shaped objects, measuring from 20 to 100 nanometers, resemble fossils of bacteria found on Earth, but are much smaller. A nanometer is one billionth of a meter. Most of the objects, then, are about 0.001 as wide as a human hair. In early 2001, additional evidence in the meteorite was suggested by microscopic crystals, arranged in long chains, which could have been formed only by living creatures. There are scientists who take the life-on-Mars theory up another notch. They suggest the possibility that life on Mars, or elsewhere in the universe, could have predated life on Earth and that the seeds of life on Earth were carried here by a meteor, a highly controversial theory known as *panspermia*. That is a concept with extraordinary implications, and faces tremendous skepticism. The odds of a biological hitchhiker surviving being blasted into space and weathering the heat and radiation before crashing into Earth are beyond astronomical.

When the first findings about the rock were announced in 1996, tabloid newspapers picked up the story and ran bold headlines about life on Mars. NASA announced that it would give samples of the meteorite to other scientists throughout the world so they could help verify or disprove the claim. Scientists plan to test the meteorite for further evidence of life because the evidence so far is inconclusive. Plenty of skeptics dismiss the shapes in the rock as unreliable, circumstantial evidence. But the hunk of basalt found in Antarctica provided enough shreds of possible evidence to add up to the intriguing possibility of ancient life on Mars. Most scientists believe that there will be no definitive answer to the life-on-Mars question until probes return to the planet, or people are sent to Mars.

Will we ever go to Mars?

The hints of water and past life make Mars the prime contender for future human exploration. Setting aside science fiction for the moment, it is a task of enormous complexity, danger, and cost.

Theoretically, one way to do that is a process called *terraforming*, which would attempt to make the Martian environment more like Earth's. There are serious advocates of space exploration, such as the Mars Society, an organization of four thousand people intent on lobbying for colonizing Mars, who envision a plan to create a "greenhouse effect" on Mars—raising the planet's temperature and melting the ice at the poles. A technologically daunting and ambitious project, it would essentially mean creating the conditions that two hundred years of burning fossil fuels on Earth has created—a cloud of gas over the planet, which would trap the heat of the Sun and raise the planet's temperature. Ironically, it is exactly this process that, many scientists argue, we must reverse if we are to prevent an ecological disaster on Earth, as worldwide temperatures rise and glacial ice melts, raising sea levels around the world. Terraforming Mars is a scheme fraught with technical difficulties—if not impossibilities—and is not being seriously considered.

The first problem to solve is getting astronauts to and from Mars—a very long road trip. The 470 million miles that the Mars Polar Lander had to cross to reach Mars is about two thousand times farther than the distance between the Earth and the Moon. With current technology and proper timing that would close the distance between the Earth and Mars, a Mars trip would take at least six months. Besides distance, there are such imponderable hazards as higher radiation levels, meteorites, and potential malfunctions that a crew could not handle. Given the spotty track record of Mars missions past, the pursuit of a manned mission to Mars is clearly in the drawing board-stages only. As of 2001, NASA had no active plans to put people on Mars and with the budgetary constraints NASA was beginning to confront in early 2001, such planning seems unlikely. In April 2001, the *Odyssey* spacecraft rocketed off on the 286-million-mile journey to Mars, the first attempt to reach Mars since those humbling failures in 1999. After a six-month flight, *Odyssey* will spend two and a half years on a geological survey, searching for water on or beneath the Martian surface—water that would hold the promise of life. There are six planned robotic missions to Mars planned for the next decade. A mission, scheduled for 2008, is designed to land on Mars and then return to Earth with samples of Martian rock and soil.

MARS: VITAL STATISTICS

Diameter: 4,223 miles (6,796 kilometers)

Mean distance from the Sun: 141,600,000 miles (227,900,000 kilometers)

Distance from the Earth:

  shortest—34,600,000 miles (55,700,000 kilometers)

  greatest—248,000,000 miles (399,000,000 kilometers)

Length of year: 687 Earth days

Rotation period: 24 hours, 37 minutes

Temperature: −225°F to 63°F (−143°C to 17°C)

Atmosphere: carbon dioxide, nitrogen, argon, oxygen, carbon monoxide, neon, krypton, xenon, water vapor

Composition: silicon, iron

Number of known satellites: 2
- Phobos
- Deimos

Around 2.5 billion years old—about half the age of the solar system—the two moons of Mars must have been formed in a collision after the solar system's birth. They are probably small *asteroids* caught by the gravitational pull of Mars.

Did an asteroid kill the dinosaurs?

Next stop on the tour is really a dodge of sorts because it is not another planet. Out between the orbits of Mars and Jupiter is a broad band of space known as the Asteroid, or Main, Belt. The solar system contains millions of rocks—actually, minor planets, since they orbit the Sun—called asteroids, which is from the Greek for "starlike." But there is really nothing starlike about asteroids. They are bits and pieces of rock, ranging from the size of pebbles to boulders or massive rocks like Ceres, which is about 584 miles (940 kilometers) wide. Composed of rock and iron, they are though to be Creation's leftovers—fragments from the formation of the solar system. Ninety percent of asteroids exist in the Asteroid Belt, and orbits have been calculated for about seven thousand of them. The other ten percent are found in other parts of the solar system.

Like planets, asteroids rotate as they orbit the Sun and take between three and six years to complete one orbit. Some asteroids are in orbits outside the belt, such as a group that follows the planet Jupiter (see below). Some stray from the Asteroid Belt to cross paths of the inner planets. Another group, called Near Earth Asteroids, orbit the inner solar system and occasionally cross the paths of Mars and Earth. And that's why asteroids are so intriguing—and potentially threatening—to us.

Off the coast of the Yucatan Peninsula in Mexico, underneath the sea, lies a crater called Cicxulub (pronounced Cheek-shoo-loob). With a diameter of about 190 miles (300 kilometers), this crater is believed by many scientists to be the result of the impact left when a 10-mile-wide object from space came crashing into Earth 65 million years ago. The resulting explosion would have triggered *tsunamis* (tidal waves) 300 feet high and set fires that burned whole continents, producing thick clouds of dust and smoke. Blocking sunlight for months or even years, the dust cloud would have produced an effect called "nuclear winter"—killing off vegetation and interrupting the food chain. It is this massive collision that is believed to be the chief cause of the mass extinction of the dinosaurs.

The dinosaurs may be the most famous victims of an asteroid, but recent discoveries suggest that they weren't the only ones. There have been at least five major mass extinctions in Earth's history, with the dinosaurs being the most recent and most noteworthy. Based on a worldwide fossil record, it is now suspected that some 250 million years ago, before the dinosaurs existed, more than three fourths of all species perished. The disappearance of this life is believed to be one reason why the dinosaurs, which evolved from lizards, were able to eventually dominate the planet.

While now widely accepted, the cosmic collision theory is not unanimously accepted by the experts. There are serious skeptics who doubt that this disaster could completely account for the dinosaur extinction. But a different question—What was it that hit the Earth?—has begun to attract more attention. Debate has raged in scientific circles between a rocky asteroid, or a loose, sooty ice ball—a comet (see below). One piece of recent evidence points to an asteroid being the culprit. A UCLA geochemist, Frank Kyte, reported finding a piece of rock in ocean sediment taken from the Pacific Ocean. This sediment contained high levels of iridium—a mineral typical of the layer of Earth laid down at the time of the end of the Cretaceous period, 65 million years ago, the Dinosaur Doomsday. Buried within the rock was a smaller piece of rock whose texture and chemical composition indicate that it is a fragment of an asteroid rather than a comet

There is even a bigger question still. Could it happen again? Hollywood thinks so, as two recent science-fiction disaster films take on that question. In the box-office hit *Armageddon*, a group of astronauts is sent aboard the space shuttle on a likely suicide mission to deflect an oncoming asteroid. And in *Deep Impact*, the actual aftermath of a

"All I'm saying is __now__ is the time to develop the technology to deflect an asteroid."

so-called "extinction event" is shown. Huge tsunamis inundate coastal areas such as Washington, D.C., and New York City, fiery chunks of debris land with the impact of nuclear bombs, and a sky blackened with smoke and ash cuts off the Sun.

So how likely is it that the Earth will be struck again? And what could we do about it? Some of the answers may come from a major NASA project launched in 1996, which started to bear fruit in 2000. A spacecraft called the NEAR-Shoemaker was designed specifically to study asteroids. Its name stands for Near Earth Asteroid Rendezvous; the Shoemaker designation was added to honor Gene Shoemaker, a geologist who influenced decades of research on asteroids but died in an automobile accident in 1997. About the size of a car, the NEAR craft is about nine feet long, with four solar panels to provide power.

NEAR conducted the first long-term, close-up study of an asteroid named Eros, one of the largest near-Earth asteroids. "Near Earth" is a relative term. Eros is more than 196 million miles (316 million kilometers) away. In a clever publicity ploy, NASA announced NEAR's contact with Eros, named for the Roman god of love, on Valentine's Day, 2000.

Eros was chosen as the target of the celestial love match because it is large. Looking very much like an Idaho potato in space, Eros is about 21 miles (33 kilometers) long, 8 miles (13 kilometers) wide, and about that thick. Most of the identified near-Earth asteroids are a little more than a half mile (about a kilometer) across. Eros is one of the Near Earth asteroids that come within 121 million miles (195 million kilometers) of Earth. That may sound like a lot of breathing space, but it is a relatively small distance in space terms. After a few months in orbit around Eros, NEAR had provided highly detailed views of the asteroid's features and geological composition, evidence to explain the origins of asteroids. One year after its rendezvous with Eros, in February 2001, NEAR descended and made a perfect landing on Eros. For two weeks, NEAR beamed back quality data about the space rock's surface composition. By the end of February 2001, NASA announced it would break its links with NEAR, which was one of the earliest and most successful of a new generation of NASA spacecraft in the agency's Discovery program that focuses on probes that are aimed to be "cheaper, faster, and better" than NASA's previous generation of space exploits.

While Eros posed no threat to Earth, one of the NEAR-Shoemaker's key goals is learning more about the possibility of an asteroid impact and how to detect other asteroids on a potential collision course with our planet.

And if one is discovered, what could we do? There are basically two theoretical approaches to saving the planet from a collision with an Earth-bound asteroid. Both dangerous and technically difficult, they involve using rockets to alter or destroy the asteroid. The first option is to use our nuclear arsenal to blow up the incoming asteroid. The disadvantage to this strategy is the potential downside of creating an awful lot of radioactive debris that might rain down on our heads. The preferred approach would be to deflect the asteroid by using a rocket to nudge it out of the way and alter its course, deflecting the asteroid away from Earth, perhaps even by landing a rocket booster on the asteroid and turning it into a spacecraft we could actually control.

One of the NEAR craft's goals is to determine just where the center of Eros is, because any push would have to be aimed at the asteroid's center. Determining the asteroid's composition and density is also important because there would be a danger that any push that is too hard might break up a fragile asteroid, creating many smaller, but still potentially dangerous, pieces of debris that could reach Earth.

This isn't just the stuff of science fiction. As the Yucatan crater or the Barringer Meteor Crater in Arizona show, Earth has been the target of many space rocks over the ages, some small, some large. In March 1989, an asteroid passed within 414,000 miles (690,000 kilometers) of Earth. It was largely undetected because the Moon was so bright at the time that the asteroid was lost in the glare. In another warning, two astronomers at the Jet Propulsion Laboratory computed orbits of known asteroids and calculated that about one hundred asteroids will come within 24 million miles (40 million kilometers) of Earth, and each year astronomers discover new asteroids coming very close to Earth. In 1994, the closest call came when 1994 XL1 came within 65,000 miles (about 100,000 kilometers) of Earth—four times closer than the Moon.

One problem is that we often can't see asteroids until the "last minute" because they are so small, and even the smallest of asteroids could wreak tremendous havoc on Earth. An asteroid measuring only 82 yards (75 meters) across could destroy major metropolitan areas. A 383-yard-wide (350-meter) asteroid could destroy an area the size of a small state or produce catastrophic tsunamis in an ocean impact. An asteroid of 4 miles (7 kilometers) would mean global mass extinction and long-term climate change.

The good news is that after scientists' ignoring this problem or dismissing it as unrealistic, asteroids are being tracked by an increasing number of observatories, and the search techniques have improved. More good news came in early 2001 when NASA scientists taking a census of large asteroids—of at least one kilometer (0.6 mile) in diameter—cut their estimate of such asteroids in our immediate neighborhood in half, down to between five hundred and one thousand. Yet, the threat, while small, remains real. A statement issued by the International Astronomical Union in 2001 pointed out, "As long as only a small fraction of NEOs (Near Earth Objects) has yet been detected and tracked, the danger of an impact by an unknown object remains real. History tells us that catastrophic impacts are exceedingly rare, occurring at intervals of hundreds of thousands or even millions of years. . . . Astronomers must give full and urgent attention to every new discovery, and this they do."

JUPITER

If you went to Jupiter

The fifth planet from the Sun, Jupiter is named for the king of the Roman gods, a fitting name for what is by far the largest planet in the solar system. Its mass equals 70 percent of all the other planets combined—318 times the mass of Earth. More than 1,300 Earths could fit inside it. An icy layering of clouds shrouds Jupiter's surface—mostly hot liquid hydrogen, probably with a rocky core larger than Earth. Jupiter has a complex weather system, which generates massive bands of clouds that swirl across its surface and also includes the planet's best-known feature, the Great Red Spot, an enormous cloud of gases. Like its neighbors in the outer solar system, Jupiter has three thin rings around its equator. They are much fainter than the more famous rings of Saturn (see below). Jupiter's rings appear to consist mostly of fine dust particles. After the Moon and Venus, Jupiter is usually the brightest object in the night sky. At the moment, only Mars is sometimes brighter when at its closest point to Earth. But, by 2004, the International Space Station will outshine both Jupiter and Mars.

What is Jupiter's Great Red Spot?

Although it sounds like a teenager's worst blemish nightmare, the Great Red Spot is considerably more impressive than the ugliest overnight pimple on the end of the nose.

Jupiter's most outstanding surface feature is the Great Red Spot, a swirling mass of gas resembling a hurricane, first seen in 1665 by the Italian-French astronomer Giovanni Cassini. The edge of the Great

Red Spot circulates at a speed of about 225 miles (360 kilometers) per hour, but scientists are still not certain what keeps this great storm growing. The spot remains at the same distance from the equator but drifts slowly east and west. Yet, this isn't just any old storm—the diameter of the Great Red Spot at its widest is about three times the diameter of Earth. Although usually red, the color of the spot varies from brick-red to slightly brown, and sometimes to gray and white. Its color may be due to small amounts of sulfur and phosphorus in the ammonia crystals, or a chemical phosphine that turns red in the presence of sunlight.

Could Jupiter be an underachieving star at the center of a solar system that fizzled?

Because Jupiter is so large and has so many satellites orbiting it, many scientists have questioned whether Jupiter could be a star that never

"*The temperature on Jupiter today is minus two hundred degrees, with the atmosphere unusually heavy in ammonia and methane.*"

made the grade. Jupiter radiates more energy than it receives from the Sun—almost twice as much. But unlike the nuclear furnace of a star, Jupiter gives out heat because it is slowly shrinking under the pressure of its own gravity. This causes the gas giant to compress and heat up. Scientists have wondered whether Jupiter could have been much hotter in its early history. But most believe that to generate the pressure needed to trigger the nuclear reactions that make stars shine, Jupiter would have to be eighty to one hundred times its present size.

Apart from Jupiter's great size, another reason for this speculation is the fact that Jupiter has sixteen known satellites. The four largest, in order of their distance from Jupiter, are Io, Europa, Ganymede, and Callisto. These four moons are called the Galilean satellites because the Italian astronomer Galileo discovered them in 1610 with one of the earliest telescopes in one of the most significant discoveries in the history of astronomy. To Galileo, these satellites rotating around a mother planet provided the evidence for Copernicus's idea that not all the stars were fixed and rotating around a stationary Earth.

- Io has many active volcanoes, which produce gases containing sulfur. The yellow-orange surface of Io probably consists largely of solid sulfur that was deposited by the eruptions.

- Europa ranks as the smallest of the Galilean satellites, with a diameter of 1,950 miles (3,138 kilometers). Europa has a smooth, cracked, icy surface.

- Ganymede is the largest Galilean satellite, bigger than the planet Mercury, with a diameter of 3,273 miles (5,268 kilometers). Discovered in 1610 by Galileo, it was named by German astronomer Simon Marius. In mythology, Ganymede was a beautiful young boy, the youngest son of King Tros of Troy. Zeus (the Greek counterpart of Jupiter) fell in love with the boy when he saw Ganymede tending the flocks, and abducted him. He was carried away by an eagle to Mount Olympus, home of the gods, where he served as cup-bearer to the gods.

- Callisto, with a diameter of 2,986 miles (4,806 kilometers), is slightly smaller than Mercury. Ganymede and Callisto appear to consist of ice and some rocky material, and both satellites have many craters. Callisto has a double spot in space. In mythology, she was a nymph, also much admired by Zeus. In a fit of jealousy, Zeus's wife, Hera, turned Callisto into a bear. Zeus then set the bear in the sky in the form of the constellation Ursa Major—the Great Bear.

Jupiter's remaining twelve satellites are much smaller than the Galilean moons. Amalthea and Himalia are the largest of the twelve. Potato-shaped Amalthea is about 168 miles (270 kilometers) in its long dimension. Himalia is about 116 miles (186 kilometers) in diameter. The smaller satellites were discovered from 1892 to 1974 by astronomers using large telescopes on Earth, and in 1979 by scientists who studied pictures taken by the *Voyager* spacecraft.

JUPITER: VITAL STATISTICS

Diameter: 88,846 miles (142,984 kilometers)

Mean distance from the Sun: 483,600,000 miles (778,400,000 kilometers)

Distance from the Earth:

   shortest—390,700,000 miles (628,760,000 kilometers)

   greatest—600,000,000 miles (970,000,000 kilometers)

Length of year: about 12 Earth years

Rotation period: 9 hours, 55 minutes

Average temperature: about −250°F (−157°C)

Atmosphere: hydrogen, helium, methane, ammonia, carbon monoxide, acetylene, phosphine, water vapor

Composition: hydrogen, helium, methane

Number of known satellites: 16

Voices of the Universe:
John Keats (1795–1821) from "Hyperion" (*Poems*, 1820)

Deep in the shady sadness of a vale
Far sunken from the healthy breath of morn
Far from the fiery noon, and eve's one star,
Sat gray hair'd Saturn, quiet as a stone.

SATURN

If you went to Saturn

Sixth planet from the Sun, Saturn is the second largest planet; only Jupiter is larger. Named for the Roman god of agriculture, Saturn is most famous for its gleaming rings, which make the planet one of the most beautiful objects in the solar system.

The farthest planet from the Earth that the ancient astronomers knew about, Saturn can be seen with the unaided eye. Most scientists believe Saturn is a giant ball of gas that has no solid surface but it may have a hot, solid inner core of iron and rocky material. Around this dense central part is an outer core that probably consists of ammonia, methane, and water. Above this layer lies a region composed of hydrogen and helium in a syrupy form. The hydrogen and helium become gaseous near the planet's surface and merge with its atmosphere, which consists of swirling clouds, probably of frozen ammonia. Scientists doubt that any form of life as we know it could exist on the planet.

Voices of the Universe:
Galileo, writing in code to Johannes Kepler, 1610

SMAISMRMILMEPOETALEUMIBVNENUGTTAVIRAS

Can't figure it out? Don't feel bad. Neither could Kepler, one of the greatest geniuses in the history of science. It is an anagram that mixes the letters that spell, in Latin, "*Altissimum Planetam Tergeminum*" ("I have observed the farthest planet as a triple star.")

When he first spotted Saturn, Galileo thought it had "handles." Then he believed that it was not a single star (or planet) but three together. Certain that Galileo had made an important discovery, Kepler was frustrated that he could not solve the puzzle. When Kepler told the Austrian emperor Rudolph II about the message, a royal request for an explanation was dispatched to Galileo and he responded, "Saturn is not a single star, but three together which touch each other." Working with a telescope with insufficient magnification, Galileo could not confirm his theory and later had doubts about what he had seen.

What are Saturn's rings?

Galileo could not see the rings clearly with his small telescope. In 1656, after using a more powerful telescope, Christiaan Huygens, a Dutch astronomer, described a "thin, flat" ring around Saturn. Huygens thought the ring was a sheet of some solid material.

But, in 1675, Italian-born astronomer Gian Domenico Cassini announced that he had discovered two separate rings made up of swarms of satellites. Like many early astronomers, Cassini was first hired to work in the observatory of a wealthy nobleman to do astrological predictions. To assure that his divinations were accurate, Cassini had been equipped with the finest instruments of the day. What his boss did not know was that Cassini was intent on proving that astrology was nonsense. In 1669, Cassini was invited to come to Paris to work in the new Royal Observatory by one of the ministers of Louis XIV. He accepted the job, adopted the French version of his Italian name (Jean-Dominique), and led the new observatory, which was well supported by the monarch known as the Sun King. For that obvious reason, Louis XIV sponsored serious and intense interest in astronomy and the sciences. For all of his other notable excesses as France's longest-reigning and most extravagant monarch, Louis XIV spent lavishly to ensure that his Royal Observatory would be as splendid as his other undertakings, such as the palace at Versailles. It was at the Royal Observatory of Louis XIV that Cassini made his discovery about Saturn.

With the benefit of improved telescopes, we know now that Saturn has seven thin, flat rings that surround the planet at its equator but do not touch Saturn. Saturn's rings consist of thousands of narrow ringlets, composed of billions of pieces of ice particles that orbit the planet. These pieces range from the size of dust to chunks of ice that measure more than 10 feet (3 meters) in diameter.

Saturn's major rings are extremely wide. The outermost ring, for example, may measure as much as 180,000 miles (300,000 kilometers) across. However, the rings of Saturn are so thin that they cannot be seen when they are in direct line with the Earth. They vary in thickness from about 660 to 9,800 feet (200 to 3,000 meters). A space separates the rings from one another. Each of these gaps is about 2,000 miles (3,200 kilometers) or more in width.

Jupiter, Neptune, and Uranus are the only other planets known to have rings, but their rings are much fainter than those around Saturn.

SATURN: VITAL STATISTICS

Diameter: 74,898 miles (120,536 kilometers)

Mean distance from the Sun: 888,200,000 miles (1,429,400,000 kilometers)

Distance from the Earth:

shortest — 762,700,000 miles (1,277,400,000 kilometers)

greatest — 1,030,000,000 miles (1,658,000,000 kilometers)

Length of year: about 29.5 Earth years

Rotation period: 10 hours, 39 minutes

Temperature: –288°F (–178°C)

Atmosphere: hydrogen, helium, methane, ammonia, phosphine

Composition: we believe there is a small core of rock and iron, encased in ice and topped by a deep layer of liquid nitrogen

Number of known satellites: 22. Until recently, we counted 18 moons for Saturn. But in October 2000, astronomers

announced the discovery of 4 new moons orbiting Saturn.
Little is known about them except that they are tiny and
orbit more than 9 million miles (15 million kilometers)
from the planet's surface.

Voices of the Universe:
Henry David Thoreau (1817–1862) from *Journal*, 1841

A slight sound at evening lifts me up by the ears, and makes
life seem inexpressibly serene and grand. It may be in
Uranus, or it may be in the shutter.

URANUS

If you went to Uranus

Okay, okay. It is the planet that gets the junior high school boys to
laugh. Yes, it is called Uranus and we all remember the scene in *E.T.*
in which the boys tease the young hero, Elliot, about, "Your anus. Get
it? Your anus." So let's get that particular joke out of the way.

The seventh planet from the Sun, Uranus is a giant ball of gas and
liquid. Its diameter is more than four times that of the Earth and what
we can see of the surface consists of blue-green clouds made up of
tiny crystals of methane, frozen out of the planet's atmosphere. Far
below the visible clouds are probably thicker cloud layers made up of
liquid water and crystals of ammonia ice. Deeper still—about 4,700
miles (7,500 kilometers) below the visible cloud tops—may be an
ocean of liquid water containing dissolved ammonia. At the very center of the planet there may be a rocky core about the size of the Earth.
Given these conditions, there is little chance that Uranus could support any form of life.

Unique in the solar system for its rotation, Uranus spins on a
tipped-over axis like some celestial Ferris wheel.

Images of Uranus taken by *Voyager* 2 and processed for high contrast by computers show very faint bands within the clouds parallel to
the equator. These bands are made up of different concentrations of

smog produced as sunlight breaks down methane gas. In addition, there are a few small spots on the planet's surface. These spots probably are violently swirling masses of gas resembling a hurricane about the center.

Why isn't Uranus called George?

The seventh planet from the Sun, Uranus was not discovered until 1781 by Wilhelm Herschel, a German-born church organist and piano teacher who learned music as a military-band member during the Seven Years' War. When he discovered that the oboe didn't make a great weapon, young Herschel beat a hasty retreat for England where he became William. While playing organ in a church and accompanying his sister Caroline, a singer, in the spa town of Bath, Herschel got "astronomy fever." An amateur astronomer, Herschel realized he could make his own telescopes better than those he could buy, and he became a master lens grinder whose telescope tubes were constructed like fine musical instruments. He became so good at grinding lenses that he and his sister Caroline Herschel—whose contribution to astronomy went well beyond bringing her brother sandwiches—soon devoted their lives to making telescopes. After years of shivering in the chill nights and early mornings, a cloak over his head to block out unwanted nearby light, Herschel struck astronomical pay dirt on the night of March 13, 1781. He found what he understood to be a planet, a faint greenish speck. Just as Galileo had named stars for his patrons the Medicis, Herschel decided to name his discovery "Georgium Sidus" ("George's Star") after a powerful man—King George III of England—like Herschel, another transplanted German. We don't know if George III, who was going through a bit of a manic period at the moment, felt that getting a planet named for him was rough justice for losing another piece of real estate—the American colonies—at about the same time. But he did appoint Herschel his royal astronomer with an annual salary. So Herschel and his sister Caroline, who also received a royal stipend, in essence becoming the first female professional astronomer, were able to quit their day jobs of piano lessons and singing gigs.

But European astronomers of the day rejected naming a planet for England's ruler, opting for Uranus, the first sky god and father of Saturn in Roman mythology. The reason that it took so long to discover Uranus? It is a long, long way away. Nearly two billion miles from the Sun. It is twice as far out as the sixth planet, Saturn. With his discovery, Herschel had doubled the known solar system.

Although recently surpassed in the "most moons" category by Saturn with its twenty-two, Uranus has twenty-one known satellites. Astronomers discovered the five largest satellites between 1787 and 1948. Photographs taken by *Voyager 2* in 1986 revealed an additional ten. Astronomers used an Earth-based telescope to discover two more satellites in 1997, and four more have been discovered recently.

The five visible from the Earth are named for Shakespearean characters, as are more recent discoveries such as Puck, a small moon.

- Miranda: the smallest of these five large satellites has certain surface features that are unlike any other formation in the solar system. These are three oddly shaped regions called "ovoids." Each ovoid is 120 to 190 miles (200 to 300 kilometers) across. The outer areas of each ovoid resemble a race track, with parallel ridges and canyons wrapped about the center. In the center, however, ridges and canyons crisscross one another randomly.

- Ariel

- Umbiel

- Titania: the largest, with a diameter of 981 miles (1,578 kilometers)

- Oberon

URANUS: VITAL STATISTICS

Diameter: 31,763 miles (51,118 kilometers)

Mean distance from the Sun: 1.8 billion miles (2.9 billion kilometers)

Distance from the Earth:

    shortest — 1,607,000,000 miles (2,587,000,000 kilometers)

    greatest — 1,961,000,000 miles (3,156,000,000 kilometers)

Length of year: 30,685 Earth days (84 Earth years)

Rotation period: 17 hours, 8 minutes

Temperature: –357°F (–216°C)

Atmosphere: hydrogen, helium, methane

Composition: primarily composed of hydrogen and helium but may also contain heavier elements

Number of known satellites: 21

NEPTUNE

If you went to Neptune

Remote Neptune, at times the farthest planet, more than thirty times Earth's distance from the Sun, was unknown until a century and a half ago. Its year is so long that it still has not completed a full orbit of the Sun since its discovery in 1846. Until *Voyager 2*'s fly-by in 1989, we knew almost nothing about Neptune, named for the Roman god of the sea. Although *Voyager* established that the planet shares many family characteristics with Jupiter, Saturn, and Uranus, Neptune has yet to give up its greatest secret: the source of the heat that rises from the planet's center to drive violent storms in its atmosphere.

One of the two planets that cannot be seen without a telescope, Neptune is usually the eighth planet. But, every 248 years, Neptune and Pluto trade places, so to speak. Pluto moves inside Neptune's orbit for about a twenty-year period, during which it is closer to the Sun than Neptune. Pluto crossed Neptune's orbit on January 23, 1979, and remained within it until March 1999.

Scientists believe that Neptune is made up chiefly of hydrogen, helium, water, and silicates. Silicates are the minerals that make up most of Earth's rocky crust, though Neptune does not have a solid surface like the Earth. Thick clouds cover Neptune's surface. The interior of the planet begins with a region of heavily compressed gases. Deep in the interior, these gases blend into a liquid layer that surrounds the planet's central core of rock and ice. The tilt of Neptune's axis causes the sun to heat the planet's northern and southern halves alternately, resulting in seasons and temperature changes.

In 1989, the *Voyager 2* spacecraft found that Neptune had a dark area made up of violently swirling masses of gas resembling a hurricane, and that Neptune has the fastest winds in the solar system; some are thought to blow at over 1,500 miles (2,500 kilometers) per hour. By comparison, the fastest wind recorded in the United States was 231 miles (372 kilometers) per hour on Mt. Washington, New Hampshire, and the top wind speed of the most violent hurricane is 155 miles (about 250 kilometers) per hour. This area, called the Great Dark Spot, was similar to the Great Red Spot on Jupiter. But, in 1994, the Hubble Space Telescope found that the Great Dark Spot had vanished.

How did they find Neptune without a telescope?

Neptune was the first planet to be discovered with mathematics rather than observation. After the discovery of Uranus, astronomers noticed that this planet, which they thought was the most distant planet, didn't always behave the way it should. They speculated that the gravitational pull of some unknown planet seemed to be influencing Uranus.

In 1843, John C. Adams, an English college student, set out to find the unknown planet. Adams predicted the planet would be about 1 billion miles (1.6 billion kilometers) farther from the Sun than Uranus. He completed his remarkably accurate work in September 1845. Adams sent it to Sir George B. Airy, the astronomer royal of England, but the society members didn't take the young man's work seriously.

At the same time, a young French mathematician unknown to Adams, Urbain J. J. Leverrier, began working on the same astronomical puzzle. By mid-1846, Leverrier also had predicted Neptune's position. He sent his predictions, which were similar to those of Adams, to the Urania Observatory in Berlin, Germany, where Johann G. Galle, the director of the observatory, had just charted the fixed stars in the area where the planet was believed to be. On September 23, 1846, Galle and his assistant, Heinrich L. d'Arrest, found Neptune near the position predicted by Leverrier. Today, both Adams and Leverrier are credited with the discovery.

NEPTUNE: VITAL STATISTICS

Diameter: 30,800 miles (49,500 kilometers)

Mean distance from the Sun: 2,798,800,000 miles (4,504,300,000 kilometers)

Distance from the Earth:

shortest — 2,680,000,000 miles (4,313,000,000 kilometers)

greatest — 2,910,000,000 miles (4,683,000,000 kilometers)

Length of year: about 165 Earth years

Rotation period: 16 hours, 7 minutes

Temperature: $-353°F$ ($-214°C$)

Atmosphere: hydrogen, helium, methane, acetylene

Composition: hydrogen, helium, methane; an ice cap at the North Pole

Number of known satellites: 8

One of these moons, Triton, is the coldest body yet explored in the solar system, with a surface temperature of $-235°C$.

PLUTO

If you went to Pluto

The Rodney Dangerfield of planets, Pluto gets no respect. Some astronomers don't even call it a planet anymore. Small and very far away from the Earth, it would take you 9,500 years to get there driving at 55 miles (95 kilometers) per hour. The smallest of the known planets, it is just one-fifth the diameter of Earth and is smaller than the solar system's seven largest planetary satellites (moons), including Earth's Moon. In fact, Pluto is so small that it was long believed to have been an escaped moon that once orbited Neptune. While some astronomers have begun to question if Pluto should even be considered a planet, the majority of astronomical agencies still count it as one of the Solar System Nine.

Pluto's other chief quirk is that it is the outermost planet, except for brief spells when it loops inside of Neptune's orbit and comes closer to the Sun than Neptune does, making it temporarily the eighth planet from the Sun. But the two orbits do not actually cross orbital paths, so they will never collide.

In 1996, the first detailed images of Pluto's surface were taken by the Hubble Space Telescope. They show about twelve large bright or dark areas. The bright regions, which include polar caps, are probably frozen nitrogen. The dark areas may be methane frost that has been broken down chemically by ultraviolet radiation from the Sun.

Pluto orbits the Sun once every 248 years at a distance of up to 4.6 billion miles (7.4 billion kilometers). Despite the lack of data, astronomers have worked out that Pluto is an icy, rocky world and even—at times—has an atmosphere.

Pluto is about thirty-nine times as far from the Sun as the Earth is. And that distance means that astronomers know little about Pluto's size or surface conditions because it is so far from the Earth. Pluto has an estimated diameter of about 1,430 miles (2,300 kilometers), less than a fifth that of the Earth. Pluto's surface is one of the coldest places in our solar system. Apparently partly covered with frozen methane gas and with a thin atmosphere composed mostly of methane, Pluto is an unlikely candidate for supporting any form of life.

In 1978, astronomers in Flagstaff, Arizona, detected a satellite of Pluto, which they named Charon, for the Greek ferryman who rowed dead souls across the River Styx to the world of the dead, Hades, where Pluto rules. With a diameter of about 740 miles (1,190 kilometers), this moon is so similar in size to its planet that some astronomers believe they make up the solar system's only "double planet."

(To add to this far-out confusion, another object was discovered beyond Saturn in 1977 and called Chiron. Not to be mistaken for Pluto's moon Charon, Chiron is a so-called "minor planet," once possibly thought to be a tenth planet. Named for a mythical centaur, Chiron is presumed to be a leftover from the solar system's creation that was not captured as a moon by any of the larger planets.)

Who found Planet X?

It sounds a bit like a title for a bad science-fiction movie. *The Search for Planet X!* "Coming soon to a drive-in theater near you."

But, in fact, it was the search for an actual Planet X that culminated in the discovery of Pluto, the last planet found in the solar system, in 1930. The possible existence of Pluto was suggested in the nineteenth century as the explanation for slight "wobbles" in the orbit of Uranus. In 1905, Percival Lowell, an American astronomer and the man who once searched for the canals of Mars (see above), found that the force of gravity of some unknown planet seemed to be affecting the orbits of Neptune and Uranus. In 1915, he predicted the location of a new planet, and began searching for it from his observatory in Flagstaff, Arizona. He used a telescope to photograph the area of the sky where he thought the planet would be found. He died in 1916 without finding it. This "find the missing object in space" game succeeded in February 1930. The tireless search of endless photographs of the night sky continued until, in 1929, Clyde W. Tombaugh, an assistant at the Lowell Observatory, used predictions made by Lowell and other astronomers and photographed the sky with a more powerful, wide-angle telescope. In 1930, by studying rapidly flashing photographs of the same patch of sky taken at different times, Tombaugh was able to discern Pluto's image as if it had "winked" at him. Tombaugh not only found Pluto—the planet he was looking for—but also a comet, six star clusters, a super cluster of galaxies, and more than 750 asteroids. The name Pluto—the Roman god of wealth and the underworld also known as the Greek Hades—was suggested in a letter by Venetia Burney, an eleven-year-old girl, who thought the planet would be a gloomy place. The name was also secured as an honor to Percival Lowell, whose initials make the first two letters of the planet's name.

Will we ever visit Pluto?

Early in 2001, NASA announced it was looking for proposals to develop the first mission to Pluto. Although one project, the Pluto-Kuiper Express, had been planned, that work was canceled when NASA determined that development and production costs had skyrocketed unacceptably, and, in the new era of fiscal constraints on the space agency, they pulled the plug on the project being developed at the Jet Propulsion Laboratory in Pasadena, California. That organization was already reeling from the twin failures of the pair of Mars probes in 1999, as well as an earlier Mars miss.

The challenge is stiff—to reach the most distant planet in the solar system by 2015—and do it at a capped cost of $500 million. NASA opened this challenge up for wide competition, hoping to get the most for its money in an era when the agency's watchwords are not "Reach for the stars," but "Faster, better, cheaper."

The planned mission is to explore Pluto and its moon Charon before the probe heads off to study smaller icy bodies lurking in the Kuiper Belt, the vast region of the solar system beyond Pluto (see below). With the help of a nuclear-powered probe that could be launched by 2004, the craft would cross deep space for years, arriving on target by 2012.

PLUTO: VITAL STATISTICS

Diameter: 1,430 miles (2,300 kilometers)

Distance from the Sun: 3,666,200,000 miles (5,900,100,000 kilometers)

Distance from the Earth:

  shortest—2,670,000,000 miles (4,290,000,000 kilometers)

  greatest—4,670,000,000 miles (7,520,000,000 kilometers)

Length of year: about 248 Earth years

Rotation period: about 6 Earth days

Temperature: about −387°F to −369°F (−233°C to −223°C)

Atmosphere: possibly methane, nitrogen, carbon monoxide

Composition: rock, ice, mostly frozen methane

Known satellites: 1

VOICES OF THE UNIVERSE:
ROBERT BURNS (1759–1796), from *The Vision* (1786)

Misled by fancy's meteor ray,
By passion driven;
But yet the light that led astray
Was light from heaven.

"Shooting stars" and comets: What's the difference?

Just because we've seen all nine planets, don't think you're done with the solar system. Wait, it gets bigger. Maybe now you are getting the true picture of how big the solar system *really* is. Approaching the most distant reaches of our solar system, planets give way to smaller bodies, floating bits and pieces of dust, rocks, and ice, perhaps leftovers from the very creation of the solar system itself. These pieces never quite combined as other chunks and bits of rock did to form the nascent planets more than 4 billion years ago. But in spite of their smaller size, these heavenly bodies that populate the furthest reaches of the solar system can still have a startling impact on life on Earth. Throughout history, these celestial objects have dazzled and alarmed people who wondered what dangers and horrible portents these streaking lights in the sky held for them. Occasionally those streaking lights have reached Earth's atmosphere or even fallen to Earth, sometimes with spectacular and devastating consequences, like the asteroids discussed earlier.

But the question of shooting stars and comets, which have fired human imaginations for centuries, also tends to be among the most commonly confused and misunderstood of space terms. So here's a simple field guide to those remarkable heavenly lights you might see blazing across the night sky—hopefully never to land on your parked car or in your living room.

A "shooting star" or "falling star" is not a star at all, but the commonly used name for a *meteor*. A meteor is no more than a speck of dust coming into contact with the Earth's atmosphere. Seen as a flash of light in the sky, meteors appear to the casual observer to be stars falling from the heavens or shooting across the night sky. The brightest meteors are sometimes called "fireballs."

A meteor appears when a particle as small as a piece of dust or a chunk of metallic or stony matter—called a *meteoroid*—enters the Earth's atmosphere from outer space at speeds up to 45 miles (70 kilometers) per second and burns up due to friction at a height of around 6 miles (10 kilometers). As friction heats the meteoroid, it glows and creates a shining trail of gases and melted particles. Most meteoroids that cause meteors are smaller than grains of sand that glow for about a second and then disintegrate; others are about the size of a pebble. They become visible between about 40 and 75 miles (65 and 120 kilometers) above Earth and disintegrate at altitudes of 30 to 60 miles (50 to 95 kilometers).

The Earth sweeps up an estimated 16,000 tons of meteoric material every year, and millions of meteors enter the Earth's atmosphere every day.

Where do meteoroids come from?

The part of the solar system in which the Earth lies is full of the interplanetary debris—space dust—that gives rise to *sporadic meteors*, the occasional "shooting star," and the regularly scheduled streams of meteoroids that produce *meteor showers*. On any clear night, several sporadic meteors can be seen each hour. But, several times each year, Earth encounters great swaths of dust shed by *comets* (see below).

These dust storms in space give rise to a meteor shower, which appears to radiate from one particular area in the sky, the area for which the meteor shower is named. The Perseid meteor shower, for instance, which regularly produces sixty to one hundred meteors an hour around August 12 each year, appears to come from Perseus, the constellation named for the Greek hero who slew the serpent-haired Medusa. The most brilliant meteor shower ever recorded took place in 1833, when a tremendous meteor shower was seen over North America on November 12. Thousands of meteors hurtled across the sky every hour, all of which seemed to originate from the constellation of Leo. Scientists later realized that the so-called Leonid meteor shower occurred every thirty-three years and that it had been doing so since A.D. 902.

These regular meteor showers appear without fail on certain dates of the year when the Earth plows through a cloud of debris that has spread out in the wake of a comet's orbit—probably within the past few hundred years. In other words, *meteors* are spawned by comets. The person who made that connection was Italian astronomer Giovanni Schiaparelli—mentioned earlier as the man who identified *canali* (remember, "channels," not "canals") on Mars. In 1866, Schiaparelli made the connection between the appearance of the Perseid meteors and a comet named Swift-Tuttle, which had been identified four years earlier.

Voices of the Universe:
A 1951 *National Geographic* press release

Comets are the nearest thing to nothing that anything can be and still be something.

What is a comet?

Ever throw a dirty snowball at one of your friends when you were a kid? You know, when the snow gets warm and soft, and you scoop it up with some dirt and pebbles. Unlike a pure, fresh-packed snowball, those dirty snowballs were not a thing of beauty. But you threw it, and bits of pebbles and small chunks of ice would sail off as it flew toward its target. That is an earthly approximation of a comet.

An icy body orbiting the Sun, a comet is mostly made up of ice mixed with dust—although it is far larger than the snowballs you may have packed as a kid. Instead of a handful of ice, a comet is a snowball with a central nucleus a few miles across. Actually, think "iceberg," larger than the one that sent Leonardo DiCaprio to the bottom in *Titanic*. It is made of various kinds of ices and rocky dust particles that are stuck in the ice. About 80 percent of the ice is water ice, frozen carbon monoxide makes up another 15 percent, with other frozen gases contributing the balance.

When the comet approaches the Sun, the surface ices begin to vaporize. The resulting gases and the particles that were stuck in the ice fly away from the Sun, forming a cloudy atmosphere called a *coma*. The comas of some comets reach diameters of nearly 1 million miles (1.6 million kilometers). The gas and dust also stream away to form one or more tails, and some comets' tails extend to distances of 100 million miles (160 million kilometers), which is the reason they are so visible and so spectacular in the night sky.

Dust particles released from the nucleus flow to a comet's tail because sunlight pushes against them. At the same time, the solar wind—that is, rapidly moving electrically charged particles from the Sun, or plasma—interacts with the comet's gases. The solar wind pushes the gases back into a tail. It is the tail, a long trailing streamer behind the streaking light of the core that gave the comet its name. The Greek word *kometes* means "long-haired one," and the tail of the comet must have inspired early observers to imagine a woman's long locks streaming from her head. Because of these effects, comet tails always point away from the Sun. As they return to the inner solar system, comets lose some of their ice and dust, and some eventually lose all their ice. They either break up into clouds of dust or turn into objects similar to asteroids. Some of these dust particles enter Earth's atmosphere and then glow as meteors, or shooting stars.

Astronomers believe that comets formed when the planets did—about 4.6 billion years ago. The planets formed from a collection of gas, ice, rocks, and dust, and much of the ice and dust became parts of the giant outer planets of Jupiter, Saturn, Uranus, and Neptune. The leftover bits of ice and dust remained as the comets.

Astronomers classify comets as short-period comets or long-period comets, depending on how long these bodies take to orbit the Sun. Short-period comets take fewer than two hundred years to make the trip; long-period comets take two hundred years or longer. Billions of them may reside in a halo known as the "Oort Cloud" (named for Dutch astronomer Jan Hendrik Oort who suggested its existence in 1950), which exists far beyond the orbit of Pluto, on the very outer limits of our solar system. A more or less spherical cloud around the farthest reaches of the solar system, the Oort Cloud is a cosmic storehouse of dark, frozen debris. Occasionally, pieces of this debris are plucked from the Oort Cloud into orbits that take them closer to Earth. Long-period comets arrive from the Oort Cloud. Some of these sweep past the Sun once and disappear. Others revisit the inner solar system again and again, their fiery tails lighting up the night sky. Most comets swing around the Sun and then return to distant space, never to be seen again for thousands or millions of years.

But *periodic comets* have their orbits altered by the gravitational pull of the planets so that they reappear every two hundred years or less. Periodic comets are thought to come from the Kuiper Belt, a zone just beyond the planets Neptune and Pluto, usually the planet farthest from the sun. Although the idea of the Kuiper Belt was theorized in midcentury by Gerard Peter Kuiper, a Dutch-born astronomer, evidence of it was only recently discovered by astronomers David Jewitt of MIT and Jane Luu of Harvard who found the first Kuiper Belt Object (KBO) in 1992. The KBOs went undetected, because at an average distance of around 3 billion miles (5 billion kilometers) from the Sun, KBOs are very faint. Approximately as old as the solar system's 4.5 billion years, the Kuiper Belt is thought to have originated when some of the basic materials of the solar system—the rocks, water ice, and frozen gases called *planetesimals*—failed to coalesce into planets and were pulled away from the inner solar system by the giant outer planets. The Kuiper Belt exists at extraordinary distances from the Sun with its inner edge thought to be 2.8 billion miles (4.8 billion kilometers) and its outer edge stretching anywhere from 4.4 billion to

93 billion miles (about 7 billion kilometers to 155 billion kilometers).

Of the 800 or so comets whose orbits have been calculated, about 160 return to the inner solar system again and again. The one with the shortest known period is "Encke's Comet," which orbits the Sun every 3.3 years. One of the brightest and most active periodic comets, Hale-Bopp, flew past the Earth in March 1997. First spotted in 1995 by Alan Hale and Thomas Bopp, two astronomers watching the sky from separate spots in the American southwest, Hale-Bopp's dust-and-gas tail made the comet so bright that it was discovered more than 600 million miles away from the Sun, two full years before it was visible from Earth. Comet Hale-Bopp came within 122 million miles (197 million kilometers) of the Earth in March 1997, its first visit to Earth's vicinity since 2213 B.C. This was not an especially close approach for a comet, but Hale-Bopp was especially bright because its unusually large nucleus gave off a great deal of dust and gas. The nucleus was about 18 to 25 miles (30 to 40 kilometers) across.

The connection of the suicides of thirty-nine people in the Heaven's Gate cult with Hale-Bopp was a grim modern reminder of the role comets have played in human history and superstition. Though actually beautiful and majestic when seen, for some reason, comets came to be viewed as one of history's most dread portents. One ancient superstition holds that a comet appears when the devil lights his pipe and throws away the still-burning match. But since the time of Aristotle, most thinking people accepted the Greek philosopher's notion that comets were exhalations of the Earth—emissions from volcanoes or other earthly openings that sent these objects into the sky. Aristotle's belief was based on the idea that the heavens were unchanging, so that the sporadic comets must not be of the heavens. It was not until 1577 that Tycho Brahe, the Danish astronomer, undid Aristotle's cometary explanation. Brahe was able to calculate distances to stars and comets and was the first to discover that comets moved far beyond the Moon rather than existing in Earth's upper atmosphere as Aristotle had suggested.

Comets were often viewed as harbingers of bad news, especially the death of a ruler, and there are several notable historical examples where this proved unfortunately true. The Inca king Atahualpa, imprisoned by

Spanish conquistador Francisco Pizarro in 1533, learned that a comet had been seen. Knowing that his own father had died when a comet appeared, the Incan king was understandably upset. Then the omen came true. Pizarro had the last Incan king strangled to death. One of the first recorded references to a comet comes from China, some 3,500 years ago. A single sentence in a Chinese document states, "When Chieh executed his faithful counselors, a comet appeared." In *Natural History*, the Roman writer Pliny the Elder (23–79 A.D.) claimed that comets are stars that sow terror. While a comet seen at the time of Julius Caesar's death was believed to have been Julius himself, in divine form, rising to the heavens, the emperor Nero's ascension to the throne was also marked by a comet. In that case, at least for many in Rome, the terror was real—even if the comet was not to blame.

Another king whose demise was foretold by a comet was King Harold of England. His downfall came in the notable year of 1066, and we know the comet was there, because it is visible in the famed Bayeux Tapestry that depicts the Battle of Hastings, in which William the Conqueror became king of England. We even have a name for the comet that turned out to be bad news for Harold but good news for William. Today, we call it Halley's Comet.

Who was Halley?

If you've heard of only one comet, it's a good bet it is this one, particularly if you grew up in the fifties and early sixties, listening to "Rock around the Clock" by Bill Haley and the Comets. Unlike the rock 'n' roll band, in which the name Haley rhymes with "daily," Halley's Comet is pronounced "HAL-eez" and rhymes with "valleys." This bright comet, which makes an appearance near the Earth about every seventy-six years, an average lifetime, is named for the English astronomer Edmond Halley, who, it turns out, was also quite bright. Halley identified the comet, which would be linked with his name, back in 1705, when he published A *Synopsis of the Astronomy of Comets* and correctly predicted its return in 1758.

Born near London in 1656, the son of a successful soapmaker,

Halley attended Oxford starting in 1673, an age of scientific and religious revolution. Although the ideas of Copernicus, Kepler, and Galileo were being more widely discussed and taught in Protestant England than in the more Catholic countries, the mechanics of the universe were still mysterious. Fascinated by astronomy, Halley became an observer at the Greenwich Observatory, where his mentor was Royal Astronomer John Flamsteed who had begun compiling a catalog of the stars of the northern hemisphere in 1675. Halley was given the task of sailing to St. Helena, an island in the South Atlantic, later famed as Napoleon's place of exile, to begin a similar study of the stars of the southern hemisphere. The importance of accurate stellar charts was growing as England became a sea power, and the navigational value of star charts became more significant. Halley later produced the first magnetic charts of the Atlantic and Pacific Oceans, another great aid to eighteenth-century navigators.

Whether Flamsteed was simply envious of the younger man or something else was at work, Halley and Flamsteed became bitter rivals. Flamsteed apparently took a particular objection to Halley's critical attack on the biblical accuracy of the Creation as told in Genesis. Even though England was then Protestant, the literal accuracy of the Bible was upheld. The powerful older astronomer blocked Halley's appointment to a teaching post at Oxford. Halley won a large measure of revenge when he was named Flamsteed's successor as astronomer royal in 1720, following the old astronomer's death. Flamsteed's widow kept the grudge going by clearing out all of the astronomical equipment from the observatory, claiming that it had belonged to her husband.

If Halley had done nothing else by that point, he would still be remembered for cataloging the stars and the assistance he provided to his friend and colleague Isaac Newton (see part I). After encouraging Newton to publish his theories in the *Principia*, Halley became a financial "angel" who underwrote the book's publication in 1687 after the Royal Society backed out, having lost money the year before on *History of Fishes*. Halley not only paid for it but became its master marketer, sending copies to the greatest thinkers in Europe and even dispatching a simplified "Newton for Dummies" to the King.

Before Halley, most people believed that comets appeared by chance and traveled through space in no set path. But Halley found that the paths of comets seen in 1531 and 1607 were identical with the path of a 1682 comet. Halley concluded that all the observations were of a single comet traveling in a set orbit around the Sun and he predicted that the comet would reappear in 1758 and at fairly regular intervals thereafter. Although Halley died in 1742, without seeing his prophecy come true, the comet reappeared in 1758 and, soon after was named for him.

What Halley did not know was that the first sightings of this comet were made by Chinese astronomers about 240 B.C. In the year A.D. 66 the same comet was seen marking the fall of Jerusalem to the Romans. In A.D. 451, it appeared at the time of the defeat of Attila the Hun by a combined army of Romans and barbarians. A plague in A.D. 684 was blamed on this comet, and it was notably visible during the Battle of Hastings in 1066, when William of Normandy conquered England. In 1222, it was visible at the time of the death of one of France's kings.

Halley's Comet can be seen in its orbit only as it nears the Sun. On October 16, 1982, astronomers at the Palomar Observatory in California photographed the comet when it was about 1 billion miles (1.6 billion kilometers) from the Sun. The comet made its closest approach to the Sun on February 9, 1986. In March of that year, a number of unmanned spacecraft drew near the comet and collected information about its composition and the size of its core. The Earth passes through the orbit of Halley's Comet each May and October, and dust left behind by the comet enters the Earth's atmosphere and burns, producing meteor showers called the Orionids during these months.

Did you miss seeing it in 1986? Just think, you only have to make it to 2062 to get another opportunity.

VOICES OF THE UNIVERSE:
SHAKESPEARE from *Julius Caesar* (1599)

When beggars die there are no comets seen;
The heavens themselves blaze forth the death of princes.

NOTABLE COMETS

Name	First Seen	Period of Solar Orbit (in Years)
Swift-Tuttle	69 B.C.	130
Tempel-Tuttle	1366	33
Tycho Brahe's Comet	1577	unknown
Biela's Comet	1772	6.6
Encke's Comet	1786	3.3
Flaugergues	1811	3,100
Great Comet	1843	513
Great September	1882	759
Ikeya-Seki	1965	880
Bennett	1969	1,678
Kohoutek	1973	unknown
Comet West	1975	500,000
Shoemaker-Levy	1993	(before it smashed into Jupiter in July 1994, this comet was in a two-year orbit around Jupiter)
Hale-Bopp	1995	2,380
Hyakutake	1996	63,400

What happened to Comet Shoemaker-Levy 9?

Had you been looking at the night sky with a telescope on July 16, 1994, and turned your sights on the planet Jupiter, you would have witnessed a remarkable sight. In one of the most spectacular displays of pyrotechnics seen in the solar system, a comet crashed into the planet Jupiter. First discovered in March 1993 by astronomers Carolyn Shoemaker, Eugene Shoemaker, and David Levy, it was the ninth periodic comet discovered by this trio and so designated Shoemaker-Levy 9. An unstable comet, it probably once orbited the Sun as other comets do, but had been pulled by Jupiter's gravity into an orbit around that planet. When the comet was discovered, it had broken into twenty-one pieces, probably when it passed close to Jupiter.

The fragments fell while Jupiter had its "back turned" to Earth. But the planet's rotation carried the impact sites around to the visible side within half an hour. The impacts caused large explosions, probably due to the compression, heating, and rapid expansion of atmospheric gases, scattering debris over large areas, some with diameters larger than that of Earth.

VOICES OF THE UNIVERSE:
An eyewitness account of the mysterious June 30, 1908, "explosion" in Siberia

I was sitting on the porch of the house at the trading station of Vanovara at breakfast time and looking toward the north. I had just raised my axe to hoop a cask, when suddenly . . . the sky was split in two, and high above the forest the whole northern part of the sky appeared to be covered with fire. At that moment I felt a great heat as if my shirt had caught fire. . . . I wanted to pull off my shirt and throw it away, but at that moment there was a bang in the sky, and a mighty crash was heard. I was thrown on the ground . . . and for a moment I lost consciousness. . . . The crash was followed by a noise like stones falling from the sky, or guns firing. The earth trembled, and when I lay on the ground I covered my head because I was afraid the stones might hit it. At that moment when the sky opened, a hot wind, as from a cannon, blew past the huts from the north. It left its mark on the ground. . . .

What would happen if a comet hit the Earth?

Since scientists actually got to witness the impact of Shoemaker-Levy 9, it takes the question of what happens in such an event from the theoretical to the very real. If a comet similar to Shoemaker-Levy ever collided with the Earth, it would be catastrophic. The explosions alone would be devastating. They might produce a haze that would cool the atmosphere and darken the planet. If the haze lasted long enough, much of Earth's plant life could die, along with the people and animals that depend on plants. This is the theory behind the dinosaur extinction discussed earlier (see page 132).

We have a good sense of what would happen if a rocky asteroid were to hit the Earth. Many people believe that this has already happened, although it may have been a glancing blow instead of a direct hit. In 1908, something happened in the area around the Tunguska river in Siberia. Although this was a sparsely settled region, an enormous explosion was heard, and afterward, a 20-mile (32-kilometer) area of felled and scorched trees was discovered. What is clear from the impact area is that nothing actually landed at Tunguska because there is no impact crater. Despite *The X-Files* episodes crafted around the Tunguska mystery, there was no evidence of a great extraterrestrial ship landing either. Most scientists now seem to agree that most likely a comet—but possibly an asteroid—approached the Earth and skimmed the atmosphere above the area, skipping away like a stone thrown across a pond, but still inflicting tremendous damage. Truth is often stranger than fiction.

LARGEST IMPACT CRATERS: Many objects, ranging from meteorites to asteroids to possible comets, have smashed through the atmosphere to crash into Earth during the planet's four-and-half-billion-year history. But there are relatively few visible impact craters to show for all those crashes. Some are buried under the oceans. Others have been weathered away by the erosion of wind and rain. But there are a number of impressive remnants of these past impacts. These are the largest, in size order:

Location	Diameter	Age
Vredefort, South Africa	187 miles	2 billion years
Sudbury, Ontario	156 miles	1.85 billion years

Chicxulub, Mexico	125 miles	65 million years
Manicouagan Quebec	62 miles	35 million years
Acraman, Australia	56 miles	570 million years
Puchezh-Katunki, Russia	50 miles	220 million years
Kara, Northern Russia	40 miles	73 million years
Beaverhead, Montana	37 miles	600 million years

If meteors come from comets, where does a meteorite come from?

Spawned by the comets, meteors are beautiful but basically harmless. Most of the meteoroids are just dust grains from the comet's tail. Their cousins, meteorites, are a different story. When a piece of space debris is a little larger and makes it through the protective atmosphere, it has graduated to the status of meteorite. These are no longer just a pretty flash in the night sky either, but present a potential danger. Bigger and rockier than their dusty cousins, meteorites are spawned by asteroids. Some meteorites are as old as the solar system itself. Others may originate on the Moon or Mars (such as the Allen Hills rock found in Antarctica), where they were dislodged by an explosion following an asteroid impact. These pieces of rocky debris cross the Earth's path and produce spectacular fireballs as they plunge through the atmosphere. Most meteorites fall into the ocean—three quarters of Earth is covered by water, and the law of probability applies—and others fall into other inaccessible or uninhabited regions, but some do hit land, and a large number have been found in Antarctica, where they are preserved in ice.

Every day, meteoroids amounting to more than several hundred tons of material enter Earth's atmosphere. But only the largest of these ever reach the surface to become meteorites. To make it to Earth, they must be the right size to travel through the atmosphere. Those that are too small will burn up and disintegrate, while larger meteorites may explode before reaching the Earth's surface. In general, the meteorites that make a landing weigh at least four ounces, but are often a pound or more, and the largest ever found weighs 60 tons. It fell at Hoba, a farm near Grootfontein, Namibia. However, much larger bodies, such as asteroids and comets, can also strike Earth, at which point they officially become meteorites.

When large bodies such as asteroids and comets strike a planet, they not only wreak havoc, but also produce impact craters, bowl-shaped depressions that measure up to about 10 miles (25 kilometers) in diameter. More than 120 impact craters and basins, which are somewhat large and more shallow than craters, have been found on Earth. One of the most famous, the Barringer, or Meteor, Crater in Arizona, is about 4,180 feet (1,275 meters) across and 570 feet (175 meters) deep. It formed nearly fifty thousand years ago, when an iron meteorite weighing 330,000 short tons (300,000 metric tons) struck Earth. Most impact craters and basins larger than the Meteor Crater are heavily worn away or have been buried by rocks and dirt as the Earth's surface changed.

NOTABLE METEORITE CRATERS:

Aouelloul, Mauritania: About a quarter-mile wide, this crater in North Africa is judged to be more than 3 million years old.

Brent, Ontario, Canada: Almost 2.5 miles wide, this Canadian crater was formed 450 million years ago.

Clearwater Lake, Quebec, Canada: At 14 miles across, the largest crater in North America was formed 290 million years ago.

Kara, Russia: In remote northeastern Russia, a 31-mile-wide crater dated 57 million years.

Meteor Crater, Arizona: America's best-known meteorite touchdown is only three-quarters of a mile wide, but it is also one of the most recent, perhaps only 50,000 years old.

Popigay, Russia: At 62 miles, the largest known meteorite crater is dated 39 million years.

Taban Khara Obo, Mongolia: Less than a mile wide, this crater is believed to be about 30 million years old.

Does a meteorite carry no-fault insurance?

On July 25, 2000, the Associated Press reported, "Four years after a rock crashed through the windshield of his parked car, a man has learned that the rock came from outer space and is older than the Earth. The car owner, Rick Wirth, a welder from Wisconsin, was told on Monday by Paul Weiblem, a geology professor at the University of Minnesota, that his rock is a meteorite formed 4.56 billion years ago. It struck Mr. Wirth's car in his driveway on October 21, 1996."

Given that such a sizable number of cosmic debris enters our atmosphere—around 100 tons of meteorites hit the Earth every year—why don't we hear more about them? Small meteorites lose almost all of their velocity as they pass through the atmosphere. Even so, they land about as fast as a stone falling from a high building. Larger meteorites, weighing a few pounds, strike at much higher speeds, and anybody hit by one would probably be killed instantly.

But the risk is low. A very small percentage of the Earth's surface is populated. Remember, the Earth is mostly water, so the most likely spot a falling meteorite will hit is in the middle of the ocean. Then you have all of those remote uninhabited spots like the middle of deserts, Antarctica, and mountaintops. That is why you have never heard of a meteorite fatality, although old Chinese records do mention deaths caused by rocks falling from the sky. Besides the Wisconsin meteor, there have recently been some, as the airline industry likes to call them, "near misses," or, as realists put it, "near hits." In 1992, a woman in Peekskill, New York also suffered a "cosmic collision" when a twenty-seven pound space rock damaged her parked Chevy Malibu. The astronomical records don't note whether her insurance company paid out on the collision, leaving unanswered the question of whether a meteorite carries comprehensive coverage.

Apart from auto collisions, meteorites have also hit people. In 1992, a boy playing soccer in Uganda was hit in the arm by a meteorite. Although stunned, he suffered no serious injuries. The only recorded American personal injury occurred, according to *Sky & Telescope* magazine, when Mrs. Hewlett Hodges of Alabama was struck in the hip on November 30, 1953, fortunately suffering only a bruise.

And in December 1997, an explosion in Northern Ireland blamed on political terrorists turned out to be the result of a meteorite that left a 1.2-meter (4 feet) crater.

These few accounts still seem remarkable given the tonnage of space debris that constantly bombards the Earth. Estimates are that, on any given day, Earth is showered by nineteen thousand meteorites weighing more than 3.5 ounces, of which fewer than ten a year are recorded. In fact, the only known space-rock fatality of the twentieth century was a dog killed in 1911, when a shower of forty-eight space stones, which probably originated on Mars, struck the village of Nakhla, Egypt.

Speaking of the Middle East, it seems appropriate to note what is surely the most revered meteorite on Earth. The Black Stone that resides within the temple of Ka'aba (pronounced KAH-buh, also spelled Caaba) in the Moslem holy city of Mecca, is reputed to be a meteorite. The temple, the most sacred shrine in Islam, is a small, cube-shaped building with a flat roof near the center of the Great Mosque in Mecca. Muslims everywhere turn their faces toward the Ka'aba when they pray. The famous Black Stone, enclosed in a silver ring, is built into the southeast corner of the Ka'aba. According to Muslim tradition, the Ka'aba was originally built by Abraham and Ishmael (also called Ibrahim and Isma'il), and the Black Stone was given to Abraham by the angel Gabriel. Muslims today kiss the stone as the Prophet Muhammad once did. Due to its great religious significance, the stone has never been subjected to scientific investigation. Its significance may be traced back to pre-Islamic times when "black stones," presumably meteorites, were worshiped as symbols of the moon god Hubal. In the year 930, members of an Iraqi sect took the sacred Black Stone and shattered it, but the pieces were returned, sealed in pitch, and held together by silver wire. It is in this patched-together form that they are now venerated.

HEAVIEST METEORITES FOUND ON EARTH

Name/Location	Weight	Found
Hoba West, Namibia	60 tons	1920
Ahnighito, Greenland	30 tons	unknown
Bacuberito, Mexico	27 tons	1863
Mbosi, Tanzania	26 tons	1930

Agpanlik, Greenland	20 tons	unknown
Armanty, Mongolia	20 tons	1935
Chupaderos, Mexico	14 tons	known for centuries
Willamette, Oregon	14 tons	1902
Willamette, Oregon	14 tons	1902
Campo del Cielo, Argentina	13 tons	unknown
Mundrabilla, Western Australia	12 tons	unknown

Are there any other planets in our solar system?

Until the late eighteenth century, only half a dozen planets were known to exist in our solar system. Since then, another three have been found, and with vast tracts of the outer solar system still left to explore, there is always a chance that at least one more planet awaits discovery. Astronomers believe that if such a planet exists, its orbit will be far, far beyond distant Pluto. And, like Pluto, it would probably be small, because any large planet that existed within the solar system would have a measurable impact on Pluto's orbit and no such motion has been observed.

Basically, this means it's tough for a planet to play hide-and-seek, especially now that we have far more sophisticated ways of finding big things like planets. The only hint that there might be something large and round out there is that some astronomers suggest that the shape of Kuiper Belt might be influenced by gravitational effects of an unseen planet somewhere in the vast void beyond its edges. The evidence is against the existence of another Planet X—appropriately, the Roman numeral for "10," as well as the symbol for an unknown factor. But the universe can be surprising, and we simply have to wait and see.

The next decade or so of space exploration may very well clear this mystery up. Unfortunately, the best candidate to solve the mystery, the Pluto-Kuiper Express, a NASA mission to the outskirts of the solar system planned for a launch in 2001, was scrubbed. The mission was plagued by cost overruns and was one of the first victims in the midst of NASA's enforced belt-tightening under the Bush administration.

But other space-based telescopes and probes may shed some light. The Hubble Space Telescope has already photographed a planet near a pair of stars 450 light-years away, and is certainly capable of detecting one in our own solar system. The Next Generation Space Telescope (NGST), which is set to go into orbit beyond the Moon, will be even more powerful.

MILESTONES IN ASTRONOMY
1820–1894

1821 The Catholic Church lifts its ban on teaching the Copernican System.

1822 The Roman Catholic Church removes Galileo's "Dialogue Concerning the Two Chief World Systems" from the *Index* 190 years after it was first published.

1835 Halley's Comet makes its second predicted return.

1839 Harvard College Observatory is founded, the first official observatory in the United States; in 1847, a 38-cm refractor is installed there, the largest in the world at the time.

c. 1840 William Draper is the first to take daguerreotypes of the Moon.

1844 Samuel F. B. Morse uses his telegraph system to send a message from Washington to Baltimore: "What hath God wrought?"

1845 Hippolyte Fizeau and Léon Foucault take good-quality daguerreotypes of sunspots.

Michael Faraday relates magnetism to light after finding that a magnetic field affects polarization of light in crystals; he proposes that light may be waves of electromagnetism.

On September 23, 1846, Johann Galle discovers the planet Neptune using the predictions of its position by Urbain J. J. Leverrier and John C. Adams.

William Lassell discovers Triton, Neptune's large satellite.

1847 John Herschel's *Results of Observations at the Cape of Good Hope* completes the survey of the southern sky started by Edmond Halley.

Maria Mitchell discovers a comet on October 1.

1848 George Phillips Bond discovers Hyperion, the eighth moon of Saturn.

Joseph Henry projects the Sun's image onto a screen and establishes that sunspots are cooler than the surrounding surface.

1849 Hippolyte Fizeau measures the velocity of light in air by measuring the time it takes for a beam of light to pass between the teeth of a rotating gear; the light is reflected by a mirror and stopped by the next tooth of the gear; the result, 196,000 miles per second (315,000 kilometers per second), is within 5 percent of the modern accepted value.

1850 William Crach Bond takes the first clear daguerreotype of the Moon.

1851 Léon Foucault uses a pendulum suspended in a church to demonstrate the Earth's rotation.

1852 Henry Giffard builds and flies the first steam-powered dirigible.

1856 George Phillips Bond discovers that photographs of stars reveal their magnitudes.

1858 The first Atlantic telegraph cable is laid.

The first aerial photograph is taken from a balloon flying over Paris.

1859 Gustav Kirchhoff and Robert Bunsen introduce the spectroscope for chemical analyses of metals placed in a flame; they also use it to study the chemical structure of the Sun.

October 27: Kirchhoff and Bunsen announce the first measurements of frequencies of spectral lines, a powerful new way to identify elements from the light they absorb or emit.

1862   Léon Foucault measures the distance from Earth to the Sun; in *1864*, astronomers agree on 91 million miles (147 million kilometers) as the Sun's distance from the Earth; the presently accepted mean distance is 92.96 million miles (149.6 million kilometers).

1863   Sir William Huggins uses the spectra of stars to show that the same elements exist in the stars as on Earth, disproving the Greek idea that the stars are composed of something different from the elements on Earth.

1865   Jules Verne's novel, *From the Earth to the Moon*, depicts three scientists and a journalist being shot to the Moon from a great cannon situated at Cape Canaveral, Florida.

1873   James Clerk Maxwell's *Electricity and Magnetism* contains the basic laws of electromagnetism and predicts such phenomena as radio waves and pressure caused by light rays in great detail.

1877   The Bell Telephone Company is founded on July 9.

1880   Thomas Edison's first electric generating station is opened in London; two years later, the Pearl Street Station opens in New York City, bringing electric lighting to the United States.

1883   The Brooklyn Bridge, which introduces a revolutionary method of cable spinning, is dedicated on May 24.

1884   Telephone wires connect Boston to New York.

1887   The world's first mountaintop telescope is completed and installed on Mt. Hamilton near San Francisco.

William Abney develops methods to detect infrared radiation with photographs and uses them to observe the spectrum of the Sun.

1888   Johann L. E. Dreyer's a *New General Catalog of Nebula and Clusters of Stars* is published; the designation "NGC" for an astronomical object refers to Dreyer's *New General Catalog*, in which he lists 7,840 nebulas and star clusters. It is still used.

1891 Maximilian Wolf makes the first discovery of an asteroid from photographs; he discovers about five hundred more asteroids—about a third of all known—using the photographic method.

1892 Edward Emerson Barnard discovers a fifth moon of Jupiter, the first satellite of Jupiter found since Galileo; Amalthea, as the planet is named, is the last planetary satellite to be discovered without the aid of photography or space probes.

1894 Percival Lowell founds his observatory at Flagstaff, Arizona, and begins to search for a hypothetical ninth planet.

Guglielmo Marconi builds his first radio equipment; the first device rings a bell from 10 meters (30 feet) away.

PART III

WHERE NO MAN HAS GONE BEFORE

One star differeth from another star in glory.

I Corinthians 15:41

But I am as constant as the northern star,
Of whose true-fix'd and resting quality
There is no fellow in the firmament.

SHAKESPEARE, *Julius Caesar*

When I heard the learn'd astronomer,
When the proofs, the figures, were ranged in columns before me,
When I was shown the charts and diagrams, to add, divide, and
measure them,
When I sitting heard the astronomer where he lectured with much
applause in the lecture-room,
How soon unaccountable I became tired and sick,
Till rising and gliding out I wander'd off by myself,
In the mystical moist night-air, and from time to time,
Look'd up in perfect silence at the stars.

—WALT WHITMAN,
"When I Heard the Learn'd Astronomer"

What's the difference between a constellation and a galaxy?

If the stars are really in galaxies, why do we still bother with those old-fashioned constellations?

Are we in the Age of Aquarius?

If the Milky Way isn't one of the constellations, what is it? And if we're inside it, how can we see it in the sky?

Do all the galaxies look alike?

Which galaxies are nearest the Milky Way?

When is a galaxy not a galaxy?

So, galaxies look like pinwheels and disks and squashed soccer balls. Why aren't space objects shaped like bananas or sugar cubes?

What are stars and do they twinkle?

How do stars form and what makes them shine?

What is star dust?

Red giants, white dwarfs, and black holes—how do stars change and die?

Has anyone found a black hole yet?

"Beetlejuice, Beetlejuice, Beetlejuice." Who gets to name the stars?

Are there other solar systems and planets out there?

MARTHA STEWART
TAKES OVER THE UNIVERSE

Man has always dreamed of flight, of escaping Earth, of reaching for the stars and heavens. In most mythologies or religions, the gods do not take kindly to humanity overstepping the bounds. There was, for instance, the unpleasantness at the biblical Tower of Babel. Man's attempt to build a stepped pyramid to heaven ran up against an angry God who stifled that ambition by confusing the tongues of men. The Greeks had Prometheus, who made the mistake of bringing divine fire to mankind. The gods weren't too happy about that one. Then there was Daedalus, the brilliant engineer and architect of Greek myth, who devised wings made of wax and feathers. Although he warned his son, Icarus, not to fly too high or the Sun would melt the wax, the boy didn't listen to his father and ended up plunging into the ocean. Another warning about reaching for the stars.

There are undocumented accounts of a Chinese scientist named Wan Hu who tried to rig up a chair with rockets around the year 1500. He exploded with the chair, apparently becoming history's first test–pilot casualty. Around the same time, Leonardo da Vinci was sketching out wonderful ideas for helicopters, winged machines, and parachutes. Although none of these ideas ever made it from the notebook to the experimental stage, the message—in the words of a rock-n-roll classic by the Animals—was the same: "We gotta get outta this place."

Finally, a little more than a century ago, we did it. First with balloons, then with the first airplanes at the turn of the twentieth century. In a remarkable burst of human development, we went from rickety airplanes built of wood and linen to the Moon in about seventy years—about one human lifetime. But in the past two decades, we have begun to really do it—"to go where no man has gone before"— as science-fiction visionary Gene Roddenberry put it at the prelude to each episode of *Star Trek*. Unmanned probes like *Voyager I*, which became the most distant man-made object in February 1998, are taking us to the furthest reaches of the solar system. The Hubble Space Telescope, the flawed device which was launched in 1990 and then underwent a 1993 housecall-in-space and a second-maintenance tune-up in 1999, has given us new eyes on the universe, revolutionizing our thinking about the cosmos. New planets have been discovered, distant stars seen, new theories have flowered. As the International Space Station goes into full operation, and the future replacement for the Hubble telescope, the Next Generation Space Telescope (NGST), is launched in 2009, our view of the universe will expand further still, opening up unimaginable vistas and sparking discovery. And each discovery of something more distant allows us to see farther into the past. The light we see from stars is ancient. As we look out, we are really looking back, getting closer and closer to the beginning of it all.

Moving beyond our solar system, this road trip sets off to see the rest of "out there." Now we are going "interstellar," journeying among the stars, and then going "intergalactic," exploring between the galaxies, beginning with the Milky Way, the galaxy in which our solar system resides. The measurements of the solar system may have staggered both your calculator and your imagination. But, as you leave the friendly confines of this solar system, the numbers become even more boggling. The Sun is one star of many, and now we begin to contemplate the hundreds of billions and billions and billions of stars, galaxies, and galaxy clusters that pervade the seemingly endless reaches of the universe.

What's the difference between a constellation and a galaxy?

With enough imagination, on a clear night, away from the bright lights of a big city, you can pick out some of these ancient star pictures in the sky, especially when you know what you are looking for. But what the ancients did not understand was that those stars that seemed to be grouped together were actually at enormous distances from each another and were just an insignificant fraction of the true number of stars in the universe. While some of the Greek astronomers believed there were much larger numbers of stars, their notions of infinite numbers of stars were dismissed as implausible. When Giordano Bruno suggested the infinite universe, it was heresy. It took Galileo and his telescope to actually begin the process of understanding the uncountable quantities of stars. The stars of the constellations were literally a drop in the bucket—a few grains of sand on a vast beach of stars.

But the scientific realization that there are congregations of millions or billions—up to tens of trillions—of stars scattered throughout the universe, collected in galaxies, is a thoroughly modern discovery. A *galaxy* is a system of stars, dust, and gas held together by gravity. Most of the matter in the universe is concentrated into these vast bodies. Most galaxies occur in clusters containing anywhere from a few to thousands of members. The galaxy in which we live, the Milky Way, is just one among billions of others, and our solar system lies halfway to the edge of this galaxy. The word *galaxy*, comes from the Greek for "circle of milk" and had been used since the eighteenth century as a synonym for Milky Way. An astronomer back then who said "galaxy" was referring specifically to the Milky Way.

The discovery that our Milky Way galaxy was not the only galaxy dates to 1924 and is the pioneering work of one of astronomical history's most important figures, Edwin Hubble (1889–1953), for whom the orbiting telescope is named. Hubble grew up in Chicago and attended Oxford University in England as a Rhodes Scholar. He returned to the United States with a law degree, but became interested in astronomy and joined the University of Chicago's Yerkes Observatory as a researcher studying for a Ph.D. in astronomy in 1914. He volunteered for duty during World War I, served briefly in Europe, and returned to take a post at the Mount Wilson Observatory in Pasadena, California. Save for a stint during World War II at the Aberdeen Proving Grounds weapons center, Hubble remained there for the rest of his professional life, from 1919 to 1953.

At Mount Wilson, after years of studying photographs taken through telescopes, Hubble was the first to demonstrate that the universe contains star systems other than our galaxy. It was not an original idea of Hubble's; astronomers had been debating it for some time. Building on the work of Harlow Shapley, who had been the first to discern the place of the Sun in the Milky Way, not in the center of the galaxy as had been presumed, Hubble identified certain stars as being of the same type as some stars in the Milky Way, but noted that these stars were beyond the calculated boundaries of our galaxy and, therefore, a separate galaxy, or, what he called "island universes." Hubble worked out the distances to nine different galaxies—a number that astronomy has now expanded to more than a *hundred thousand million* galaxies, each containing some *hundred thousand million* stars.

Five years after Hubble figured out galaxies, he and an intriguing associate named Milton Humason, made an even more remarkable discovery. A high-school dropout who delivered supplies to the Palomar Observatory in mule trains, Humason eventually took a job as a janitor at the observatory. From there, he worked up to becoming a part-time observer and soon proved to have an uncanny ability to photograph the skies through the telescope. The unlikely pairing of the Rhodes Scholar and the mule-skinner-turned-janitor-turned-astronomer made one of the most extraordinary discoveries in history. They saw that galaxies were moving, and the farther away a galaxy was, the faster it was moving from us. The universe, according to Hubble, was not static, or fixed in size, as people had previously thought for generations, but was, in fact, expanding—the distance between galaxies was growing all the time. This idea is the fundamental underpinning of the "Big Bang" theory, the most widely accepted scientific explanation for the creation of the universe (see part V).

If the stars are really in galaxies, why do we still bother with those old-fashioned constellations?

In modern astronomy, "constellation" refers to a particular region of the sky, often enclosing the ancient "figure" that might have first been imagined thousands of years ago, but now also includes the surrounding area. The constellations are the designated regions on the *celestial sphere*—an imaginary sky grid that encloses the Earth. If you could imagine Earth encircled within a great ball or bubble, that enormous sphere would be the celestial sphere. Modern astronomy has divided this imaginary ball around the Earth into eighty-eight areas for the purposes of identifying and naming celestial objects. By charting the sky, astronomers have placed the constellations, like continents and countries, on a map. Every portion of the sky is now said to belong within a particular constellation. So, while astronomers no longer use these star pictures to divine the future, as the astronomers-astrologers of ancient times once did, they have a usable, uniform reference point for dividing the sky into measurable pieces.

The ancient constellations were simple arbitrary patterns of stars in which early people visualized gods, sacred beasts, and mythological heroes. The forty-eight constellations known at the time of Ptolemy are now known as "the ancient constellations." They were added to over the years, first with the inclusion of the southern hemisphere constellations that began to be charted between the early 1600s and 1754. The modern constellations were sometimes named for friends, patrons, or recent inventions such as the telescope and microscope.

A few of these constellations also contain smaller, identifiable shapes within them, asterisms. Probably the most familiar examples are the easily identifiable Big Dipper, an asterism within the constellation Ursa Major, and the Little Dipper, an asterism in the constellation Ursa Minor.

The concept of the constellation, once thought to have so much power over human affairs, is now a convenience, a useful tool for astronomers indicating a direction in the sky in which any space object can be found.

Since the naked eye cannot distinguish depths and distances in space, the stars that are visible all seem to be at a similar distance and on the same plane. But, of course, most of these individual stars are far apart, and the ones that are visible to the naked eye are separated by hundreds and even thousands of light-years. It is all a matter of perspective. Basically, it is a bit like looking at a map of the United States and assuming that New York City and Denver are the same size and at the same altitude because they are just two dots on a flat piece of paper. But the stars that we seemingly see against a "flat" sky are actually many different sizes and distances from Earth. For example, the stars that make up the easily identifiable asterism Big Dipper seem to be grouped closely together at approximately the same distance. In fact, most of the stars in the Big Dipper range between fifty and one hundred light-years from Earth. Dubhe, Arabic for "bear" and the star at the "spout" of the Big Dipper, is more than one hundred light-years from Earth. Alkaid, the first star in the Dipper's "handle," is more than two hundred light-years away.

Until 1930, these constellations had rather arbitrary boundaries and names. But then a group of international astronomers, the International Astronomical Union (www.iau.org), which was founded in 1919, was given legal responsibility for establishing recognized astronomical designations, and specified the names and boundaries of the eighty-eight recognized constellations.

Are we in the Age of Aquarius?

Immortalized in the rock musical *Hair*, the sixties were anointed the "Age of Aquarius," the beginning of an era of universal harmony, peace, and joy. So that didn't quite work out.

But is this really the Age of Aquarius? No. This is another confusion between the worlds of astronomy and astrology. A holdover from astrological beliefs, a so-called age is identified by the name of the constellation—the area of the sky—in which the Sun appears to be located on the first day of spring (the vernal equinox).

Back to basics. To someone on Earth, the Sun appears to move across the sky from east to west. It also appears higher or lower in the sky, depending on the season (and your location on Earth). Generally, the summer sun is higher in the sky and the winter sun is lower in the sky. And the reason for that, class? All together: "The Earth is tilted on its axis." Right? Of course, you do *really* understand that Earth is moving around the Sun. (You do, don't you?) The early astronomers, who were also very interested in astrology, charted the Sun's *apparent* path through the sky, relative to the positions of other stars, and called this imaginary path in the sky the *ecliptic*. Still with this? The ecliptic was divided into the twelve signs of the traditional zodiac. The constellation, or zodiac sign, in which the Sun rose on the vernal equinox thus determined the astrological "age."

Over a long period of time, the Sun's location on the vernal equinox changes, moving from one sign of the astrological zodiac to the next as a result of the *precession*—the wobble of Earth's axis. Currently the vernal equinox occurs in Aries. So, to be precise, we are in the "Age of Aries."

"Aquarius" might have scanned much better lyrically, but in fact, the Sun won't be in Aquarius on the vernal equinox for another six-hundred years. Maybe that explains why we missed out on all that peace and harmony.

THE CONSTELLATIONS

As we have seen, the original dozen constellations of astrology have been supplemented. Modern astronomy divides the sky into eighty-eight areas for the purpose of naming and identifying celestial objects. This list is also divided between the northern constellations and those of the southern sky.

THE FIFTY-FOUR NORTHERN CONSTELLATIONS:

These are constellations most easily visible from the northern hemisphere, although some can be seen as far as 30 degrees south of the equator. The ancient Greeks listed forty-eight constellations; the modern constellations are identified in this list by an asterisk.

Andromeda   The Princess, or Chained Maiden. Claiming that she was more beautiful than the sea nymphs, Andromeda angered Poseidon. The sea god sent Cetus, a sea monster, to ravage her country. To save their land, Andromeda's parents were told to sacrifice her. They chained her to a rock and left her to be devoured by Cetus, but she was rescued by the heroic Perseus.

Aquarius   The Water-bearer. One of the most ancient constellations, Aquarius appears as a boy spilling water from an urn. In Sumerian (ancient Babylonian) myth, this was part of the ancient deluge story that is thought to have been adapted by the later Hebrews as the biblical Flood.

Aquila   The Eagle. Another ancient constellation, Aquila was identified as a bird around 1200 B.C.

Aries   The Ram. Made of just three small stars, Aries is another of the very oldest constellations.

Auriga   The Charioteer. In Greek mythology, the inventor of the chariot.

Boötes   The Herdsman. Shaped like a kite, he is depicted in legend as leading the "hunting dogs" (below).

*Camelopardalis   Giraffe. This modern constellation was first outlined in 1624. Its Latin name derives from the fact the giraffe, in ancient times, was thought to have the head of a camel and the spots of a leopard.

Cancer   The Crab

*Canes Venatici   The Hunting Dogs. Conceived in 1687, they are often depicted as two greyhounds that pursue the "Bears" (below).

Canis Major   The Great Dog. Originally this name referred only to Sirius, the brightest star in the night sky, but now includes an entire constellation.

Canis Minor   The Little Dog. Among the smaller constellations, this consists of only three stars.

Capricornus   The Goat, or Sea Goat. This may have been the first constellation recognized in prehistoric times. Babylonian tablets from around 3000 B.C. depict the group of stars and ancient Greeks associated it with Amalthea, a goat that suckled the infant Zeus.

Cassiopeia   The Queen. In this bright configuration, the Greeks saw a seated woman they thought of as the queen and mother of Princess Andromeda.

Cepheus   The King. The Greeks identified old Cepheus as the seated ruler, husband of Cassiopeia and father of Andromeda.

Cetus   The Sea Monster, or Whale. Among the earliest constellations identified, it was associated by the Babylonians with their mythical cosmic dragon Tiamat, while the Greeks saw it as the sea monster that threatened Andromeda until it was slain by Perseus.

*Coma Berenices   Berenice's Hair. Seen as a woman's braided hair, is named for Berenice II, the queen of the Egyptian king Ptolemy III (circa 247–221 B.C.). A real person, as opposed to the mythic figures of most constellations, Berenice had vowed to cut off her beautiful long hair if her husband returned safely from a dangerous military expedition. Upon his safe return from battle, she cut her hair and placed it in the temple of Aphrodite, the goddess of beauty, and it disappeared. The court astronomer explained that the goddess had placed the plaited hair in the sky for all to see.

Corona Borealis   The Northern Crown. A U-shaped grouping resembling a crown, which the Greeks associated with the crown of Ariadne, daughter of the king of Crete. But to the Shawnee tribe of North America, it appeared as a circle of dancing maidens.

Corvus   The Crow. This legendary bird was sent to fetch a cup of water for Apollo, but was too slow and was condemned to look at The Cup (below) but never drink from it.

Crater   The Cup. Representing the cup of legend used to torment the crow, this grouping was also thought to be the cup of nectar drunk by the gods of Olympus.

Cygnus   The Swan. Also known unofficially as the Northern Cross, this is a large bright constellation in a crosslike shape that resembles a bird with wings extended in flight.

Delphinus   The Dolphin. A group depicting the animal seen throughout mythology as a friend and rescuer of men and gods.

Draco   The Dragon. With a sinuous pattern, this group of stars has stood for the mythological dragons of many cultures. One of the stars in the group, Thuban, was considered the North Star four thousand years ago, when the Egyptians oriented the pyramids towards it. Due to the wobble of the Earth's poles (*precession*) Polaris is currently the North Star.

Equuleus   The Little Horse, Foal, or Colt. A small grouping seen as a horse's head, it is not to be confused with Pegasus, the Winged Horse (below).

Eridanus The Celestial River. The second longest constellation after Hydra (below), its group was seen as the Nile River by the Egyptians and the Euphrates River by the Babylonians.

*Fornax The Furnace. An eighteenth-century creation of a French astronomer, it has only three faint stars.

Gemini The Twins. Seen as Castor and Pollux, sons of Leda and Zeus, and the brothers of Helen of Troy.

Hercules The Strongman. Named for the legendary Greek hero, this group was connected in earlier cultures with such heroes as Gilgamesh of the Sumerian epic. He was seen kneeling, club raised, with his foot on the head of Draco, the Dragon.

Hydra The Sea Serpent. The Greeks saw this group as the six-headed monster slain by Hercules.

*Lacerta The Lizard. A modern constellation, it was created by German astronomer Johannes Hevelius around 1687, when he described it as a lizard or a newt.

Leo The Lion. While it was recognized by some of the ancients from the Sumerians to the Romans as a lion, the Chinese saw it as a horse and the Incans as a puma. Leo's head is formed by an asterism known as the Sickle.

*Leo Minor Little Lion. A modern constellation, this was drawn around 1687 from a few faint stars.

Lepus The Hare, or Rabbit. From ancient times, this group was associated with the Moon, a connection based on the fertility of rabbits and the link of the lunar cycle to woman's fertility cycles.

Libra The Balance, or Scales. An ancient member of the zodiac, this group was once called the "Claw," and was viewed as part of the adjacent Scorpius. However, during Roman times, the Sun appeared in Libra around the autumnal equinox, when there is equal day and night, so the constellation has been viewed as a balance or scales ever since.

*Lynx   The Lynx. A modern group created about 1687 by Johannes Hevelius who noted the faintness of the stars by saying observers would need to be "lynx-eyed" to see these stars.

Lyra   The Lyre. To the Greeks, this was the instrument that Orpheus played to persuade the god of the underworld, Hades, to release his wife, Eurydice.

*Monoceros   The Unicorn. A modern grouping named about 1625 and based on the mythic unicorn, which goes back to the Assyrians and was probably based upon sightings of a rhinoceros.

Ophiuchus   The Serpent-bearer. One of the oldest recorded constellations, it depicts a man entwined with Serpens (below). He is identified with the legendary god of medicine, Asclepius, the "blameless physician" of Homer's *Illiad*.

Orion   The Hunter. Perhaps one of the best-known constellations, because it is so identifiable and contains many bright stars. In Greek legend, the boastful hunter Orion, a giant renowned for his good looks, was blinded for raping a princess. Orion was later given his sight back, but was then killed and placed in the sky when he tried to seduce Artemis, the cruel goddess of the hunt.

Pegasus   The Flying Horse. The winged horse was born in Greek myth when Perseus killed the Gorgon Medusa, and her blood mixed with sea foam.

Perseus   The Hero. One of the greatest heroes of Greek myth, he killed the Medusa and rescued Andromeda from the sea monster Cetus. The brilliant Perseid meteor showers (see part II) visible in mid-August appear to originate in this constellation.

Pisces   The Fish. Since earliest recorded times, these stars were identified as a pair of fish.

Piscis Austrinus   The Southern Fish. Another ancient constellation, it was commonly seen as a fish drinking from the water poured by Aquarius, the constellation directly to its north.

Sagitta   The Arrow. This alignment of stars was said to be many different arrows in myth, including the one that Apollo used to kill the Cyclops in Greek myth, and Cupid's arrow.

Sagittarius   The Archer. First identified as the Sumerian god of war, this group was seen later by the Greeks as a satyr or centaur—half-man, half-horse. While it may be hard to see the archer, it is easier to identify the asterism known as the Teapot in this grouping.

Scorpius   The Scorpion. This group was named for the scorpion that stung and killed Orion.

***Scutum**   The Shield, or Sobieski's Shield. A modern grouping, it was named for the seventeenth-century king of Poland, John III Sobieski, and was supposed to represent his coat of arms.

Serpens   The Serpent. The only constellation that is seen in two separate parts. Its head and tail are on either side of the Serpent-bearer.

***Sextans**   The Sextant is a modern grouping named by Johannes Hevelius to honor the astronomical sextant, the device used to compile the first accurate star maps.

Taurus   The Bull. One of the oldest recognized constellations, it represented strength and fertility to many ancient cultures.

Triangulum   The Triangle. A small, dim grouping was seen by the Egyptians as symbol of the Nile delta and as the triangular isle of Sicily by the Greeks and Romans.

Ursa Major   The Great Bear. After Orion, this is probably the best-known constellation. To Ancient Greeks and Native American tribes it was a large bear, and includes the bright stars that form the Big Dipper.

Ursa Minor   The Little Bear. Both a constellation and an asterism, this is the Little Dipper, which contains the current North Star, Polaris. Ursa Minor was first recognized about six hundred B.C. by Thales of Miletus.

Virgo The Virgin, or Maiden. The second largest constellation, and one of the oldest constellations recognized, it has been associated with legendary beauties throughout mythology, such as the Babylonian Ishtar, Egyptian Isis, and the Greek Athena.

*Vulpecula The Fox. Named by Johannes Hevelius in 1690. There was originally a goose in the picture, but most star charts show only the fox today.

THE SOUTHERN CONSTELLATIONS

Antlia The Air Pump. Like many modern constellations, this grouping is not actually shaped like its name, but was named in honor of the compressed air pump, a key piece in the Industrial Revolution.

Apus Bird of Paradise. This group was named for spectacular birds native to New Guinea.

Ara The Altar. This was named for the altar of the centaur Chiron, believed to be the wisest creature on Earth, who is credited with teaching mortals how to draw the lines connecting stars in constellations.

Caelum The Chisel. One of the smallest, dimmest constellation areas, it was named by the French astronomer Nicholas-Louis de Lacaille.

Carina The Keel. One of four constellations named for parts of a ship.

Centaurus The Centaur. Not to be confused with Sagittarius, the other centaur in the sky, although both are said to be the wise Chiron.

Chamaeleon The Chameleon. A faint grouping first drawn in 1603.

Circinus The Drawing Compasses. This is a fairly obscure grouping drawn in the eighteenth century.

Columba The Dove. A group first recorded as a constellation in 1679.

Corona Australis   The Southern Crown. Although Southern, it was included among Ptolemy's forty-eight original constellations.

Crux   The Southern Cross. Although it is the smallest constellation, it has a distinctive crosslike pattern.

Dorado   The Goldfish, or Dolphinfish. First registered in 1603 by Johann Bayer, it is not the goldfish of millions of households, but a seagoing fish also known as the Hawaiian mahi-mahi.

Grus   The Crane. A group first mapped in 1603.

Horologium   The Clock. Mapped in 1750, this group was named in honor of the pendulum clock invented by Galileo.

Hydrus   The Water Snake. First mapped in 1603, it should not be confused with the larger and older Hydra.

Indus   The Indian. Created in 1603, it is also known as the American Indian because it was named during the early period of the exploration of the Americas.

Lupus   The Wolf. Although one of the southern constellations known to the ancients, it is rather faint and includes no named stars.

Mensa   The Table, or Table Mountain. It was named after the location of an observatory near Cape Town, South Africa.

Microscopium   The Microscope. It was named in 1750, along with the telescope (below).

Musca   The Fly. In 1603, this group started out being described as the Bee. Then Edmond Halley called it "the Fly Bee," but it was later shortened to "the Fly."

Norma   The Level, or Rule. Another of the constellations named in 1750, referring to the carpenter's tool.

Octans   The Octant. Named in 1750 in honor of the device used by astronomers to measure the angle between the horizon and celestial objects. The device called an octant later evolved into the sextant, which has been used by sailors throughout history, right up to modern times. Today it is being made obsolete by Global Positioning Satellite (GPS) devices, which provide a much more accurate measurement of an object's location.

Pavo   The Peacock. From the 1603 atlas of Johann Bayer, this was thought to be the bird sacred to Hera, Zeus's wife.

Phoenix   The Phoenix. Another group that first appeared in Bayer's 1603 star atlas, this constellation was named after the mythic bird that was capable of being consumed in flames and then arising from the ashes as a new bird.

Pictor   The Painter. An undistinguished grouping formed in the 1750s by the French astronomer Nicholas-Louis de Lacaille, who first called it "the Painter's Easel."

Puppis   The Stern, also known as Poop. The name refers to the poop deck of a ship, which has been related to the mythological ship *Argo*, on which the Greek hero Jason sailed in search of the Golden Fleece.

Pyxis   The Compass.

Reticulum   The Net.

Sculptor   The Sculptor. This group occupies a large but otherwise unremarkable spot in the sky. Frenchman de Lacaille originally named it, "the Sculptor's Workshop."

Telescopium   The Telescope. De Lacaille formed this group by taking stars from larger nearby constellations, originally calling it "Tubus Astronomicus."

Triangulum Australe   The Southern Triangle. A small star grouping, this was first recorded in 1603 by Bayer but may have been spotted earlier.

Tucana   The Toucan. Invented by Bayer in 1603, the name of this group refers to the large-billed birds of the American tropics.

Vela   The Sails. Located in the southern sky, this is one of four constellations created by the French astronomer de Lacaille when he broke up a large constellation called Argo into several smaller groupings that include Puppis, Pyxis, and Carina.

Volans   The Flying Fish. Another constellation created by Bayer in his 1603 catalog.

VOICES OF THE UNIVERSE:
CARL SAGAN (1934–1996), *Cosmos* (1980)

A galaxy is composed of gas and dust and stars—billions upon billions of stars. Every star may be a sun to someone. Within a galaxy are stars and worlds and, it may be, a proliferation of living things and intelligent beings and spacefaring civilizations. But from afar, a galaxy reminds me more of a collection of lovely found objects—seashells, perhaps, or corals, the productions of Nature laboring for aeons in the cosmic ocean.

If the Milky Way isn't one of the constellations, what is it? And if we're inside it, how can we see it in the sky?

The Milky Way is the galaxy that includes the Sun, Earth, and the rest of our Solar System, as well as all the stars we can see. Seen from Earth, the Milky Way looks like a misty band of light spread across the night sky. Dark gaps in the band are formed by clouds of dust and gas that block out light from the stars that lie behind them. The Greeks saw this whitish cloud as a smear of milk, and that idea prevailed in the western world. But other people saw this great white cloud in other ways. Eskimos of the Arctic saw it as the snowy path of the Great Raven, their trickster god who could travel between heaven, Earth, and the sea floor—the three worlds that made up their universe. The Tartars of the Caucasus in Russia saw straw trailing behind a thief who had stolen it. In Islam, believers thought the Milky Way was the path of pilgrims on the way to the holy city of Mecca. And to the Chinese, it was the celestial river, a passageway linking Earth and heaven. The river led to the great abyss, in which the mothers of the Sun and Moon bathed their children each day before they took their places in the sky.

We now know that our Sun is only one of about 200 billion stars that make up the Milky Way galaxy. We can see somewhere in the vicinity of three thousand stars on a clear night, but this is just one *forty-millionth* of the total number of stars in the Milky Way.

Clearly, the Milky Way is huge, just like other galaxies, and astronomers use different ways of describing these enormous distances in space where miles or kilometers lose meaning. The most commonly used measurement is the *light-year*, which combines distance, speed, and time. A light-year (abbreviated ly) is the distance light travels in a year. Light travels at a speed of *186,000 miles per second* (299,792 kilometers). That works out to about 5.88 trillion miles (9.46 trillion kilometers). Let's just round it off to roughly 6 trillion miles (9.6 trillion kilometers). That's a six followed by twelve zeroes. Remember, light from the Sun takes just eight minutes to cross about 93 million miles to Earth. For even greater distances, astronomy uses a unit of distance known as a *parsec*, which equals 3.26 light-years, or a little more than 19 trillion miles (30 trillion kilometers). The key point to keep in mind is that when discussing light-years, we must realize that distance is time. When we say something like "a star is ten light-years away," it means that the light we see left that star ten years ago. That is why when we look across space, we are also looking across time. Are you feeling even less significant now? Just wait. We are just getting started on how impossibly big this galaxy is—the Milky Way is between 100,000 and 130,000 light-years in diameter.

If we could see it from another vantage point outside our galaxy, the Milky Way would look like a discus with a large bulge in the center, about 15,000 light-years across. A vast number of older stars are in the central bulge of the disk. The bulge and disk are surrounded by a sphere of stars known as a halo. The halo contains relatively old stars in dense, ball-like groups called *globular clusters*. Around the bulge, a roughly circular disk of the galaxy rotates—The Milky Way is about 10,000 light-years thick at the central bulge and much flatter toward the edges of the disk, where it is still some 2,000 light-years thick. Stars, dust, and gases fan out from the central bulge in long, curving arms that form a spiral pattern, like an enormous, whirling pinwheel. These spiral arms, marked by bright stars and glowing clouds of dust and gas called *nebulae*, are where new stars are being formed.

Obviously we can see the Milky Way from Earth, even though we are part of it, because it is so large. Ironically, we cannot be sure of the accuracy of some data that relates to the Milky Way because we are in it. The clouds of dust and gases in the Milky Way prevent us from seeing very far into the center of the galaxy, although the decade of discovery produced by the Hubble Telescope and the Cosmic Background Explorer (COBE), a satellite launched in 1988, have given astronomers valuable new information about the Milky Way.

Our solar system is located on the outskirts of the galaxy, about 25,000 light-years away from the center. The distance between the stars in our section of the Milky Way averages about 5 light-years, but stars in the galactic center are packed together much more tightly, about 100 times closer together. All stars and star clusters in the Milky Way orbit the center of the galaxy, much as the planets in our solar system orbit the Sun, which takes about 250 million years to complete one orbit of the galaxy. Almost all the bright stars in the Milky Way orbit in the same direction. For this reason, the entire galactic system appears to rotate about its center, or nucleus.

Astronomers studying radio waves and infrared rays—which can penetrate the clouds—have discovered that this central region gives off enormous amounts of energy. Studies with radio and infrared telescopes have also revealed a powerful gravitational force that seems to come from the exact center of the galaxy. Some astronomers now believe that the Milky Way's center is a massive *black hole*, an invisible object whose gravitational pull is so great that not even light can escape from it (see below).

Do all galaxies look alike?

Since Hubble's breakthrough in 1924, there have been several upheavals in the understanding of galaxies, as technology has permitted us to see and understand more about star formation. But Hubble did identify three basic types of galaxies still recognized: *spiral galaxies*, *elliptical galaxies*, and *irregular galaxies*.

- A spiral galaxy, such as the Milky Way, is a flattened, discus shaped collection of stars with a central bulge containing older stars, surrounded by a disk of younger stars. Observations show that new stars are constantly forming out of the gas and dust in these spiral arms. The Milky Way is a spiral galaxy with the Sun lying in one of its spiral arms. There is another type of spiral galaxy, described as a *barred spiral*, that has a straight bar of stars across their centers. Like a bodybuilder's barbell with circular weights at either end, the barred-spiral galaxy's arms emerge from the ends of these bars.

- Elliptical galaxies range in shape from almost perfect spheres to flattened globes. Although some are shaped like an American football, some are more rounded and others appear like a flattened soccer ball. The true shape of a galaxy cannot be determined, because we can only see it from one angle.

Elliptical galaxies contain old stars and very little gas and include the most massive galaxies known, holding as many as a trillion stars. Some elliptical galaxies may also rotate, but more slowly than do spirals. Ellipticals have much less dust and gas than do spirals, so no new stars are forming in them. Some elliptical galaxies are thought to be formed by mergers of spiral galaxies. But this merger is not what they like to call a "friendly takeover" in business. They are actually more like cosmic fender-benders, collisions between earlier generations of spiral galaxies.

A subclass of ellipticals are *radio galaxies*, which are also believed to result from galaxies in collision. While all galaxies emit radio waves—one of the forms of electromagnetic radiation that stars produce—those coming from radio galaxies are up to a million times more powerful than typical galaxies. Often, the strongest emission comes from two clouds that can extend for millions of light-years on either side of the galaxy. A candidate for the source of this enormous energy is, like the core of the Milky Way and perhaps the center of all galaxies, a massive black hole.

- Irregular galaxies, as their names suggest, have no patterns and lack a typical structure. The Large Magellanic Cloud and the Small Magellanic Cloud, visible in the southern hemisphere, are examples of an irregular galaxy. They are named for the Portuguese explorer Ferdinand Magellan (c. 1480–1521) who first described them during his attempt to sail around the globe. Magellan died in

the Philippines, but his crew completed their first successful circumnavigation of the globe and brought back his descriptions of
these "clouds," which, for centuries, no one realized were actually
other galaxies.

Which galaxies are nearest the Milky Way?

Only three of the billions of other galaxies in the universe are visible
to the naked eye. These three galaxies appear as small, hazy patches of
light. In the northern hemisphere, the Andromeda Galaxy, which is
about 2.9 million light-years away, is visible as a milky blur in the sky;
remember, at that distance, we are seeing light from Andromeda that
is nearly three million years old—light that started its journey before
there were humans on Earth. Its presence was first recorded in 964
A.D., by the Arab astronomer As-Sufi, who called it "little cloud."
Andromeda is our nearest full-size spiral galaxy, and, while similar in
structure and composition to the Milky Way, is actually much larger.
With a diameter of 150,000 light-years, it may contain 400 billion
stars, twice as many stars as the Milky Way. For centuries, astronomers
considered Andromeda to be what was then called a "nebula," a cloud
of dust and gas within the Milky Way. (The modern astronomical definition of *nebula* is the gas-and-dust cloud from which stars are born.)
It was Edwin Hubble's discovery that Andromeda actually contained
many stars that led to his landmark conclusion: Andromeda was
another galaxy, lying a great distance away.

In the southern hemisphere, two galaxies, both irregular, can be
seen. The Large Magellanic Cloud spreads over the constellations of
Dorado and Mensa. Located 169,000 light-years from Earth, it is
about a third of the diameter of the Milky Way. The Small Magellanic Cloud is in Tucana, located 180,000 light-years away, and is
about a fifth of the diameter of the Milky Way.

Discovered more recently is another nearby galaxy, called the
Sagittarius Dwarf, located a *mere* 80,000 light-years away. A "dwarf
elliptical," it is one of two miniature galaxies that are close but relatively dim, masked by their brighter neighbors or hidden behind
clouds of gas and dust. Some astronomers believe that the Milky Way

will collide with the Sagittarius Dwarf, which will then be consumed by our larger galaxy. Not to worry, though. This will not happen until sometime within the next 100 million years.

In one of the Hubble Space Telescope's recent discoveries, announced in January 2001, two galaxies were seen tethered together by a ribbon of gas and dust that was leaving one galaxy to be absorbed by the other. Presumably, the two galaxies will eventually merge completely—perhaps in another 20 million years or so.

Although these are the galaxies visible to us, they are just part of what astronomers call the "Local Group," a cluster of thirty or more galaxies, each with their own motions, yet still moving together. Galaxies are distributed unevenly in space. And while some galaxies are found alone in space, more than 60 percent of those that we know about are grouped in formations called "clusters." Clusters of galaxies range in size from a few dozen members to several thousand. This "Local Group," which has the ring of some sort of cosmic carpool, is another conglomeration of imponderable numbers. The region occupied by the Local Group is around 3 million light-years across. Beside Andromeda and the Magellanic Clouds, the group includes another major spiral system, the Triangulum galaxy, a spiral smaller than the Milky Way, which orbits the Andromeda galaxy.

That may all sound impressive, but it it still a puny cluster compared to the Virgo Cluster. Located 50 or 60 million light-years from the Milky Way, it is a *supercluster*, composed of more than *two thousand galaxies*. That is large enough to pull the Local Group toward it at a speed of nearly a million miles (about 1.7 million kilometers) per hour.

And just when you thought things couldn't get any larger, the universe throws another surprise at you. In January 2001, astronomers working at NASA's Goddard Space Flight Center announced finding a supercluster of galaxies that includes billions of stars, believed to be the largest known object in the universe. Located 6.5 billion light-years away, light from these galaxies, discovered with a telescope at the Cerro Tololo Inter-American Observatory in the Andes Mountains of Chile, began its journey when the universe was much younger, perhaps a third of its present age.

When is a galaxy not a galaxy?

Once upon a time, in the dark ages of home electronics—the 1970s—there was a heavily promoted, fancy new television called "Quasar." It sounded so high-tech, so modern, so revolutionary then. Smart marketers had picked up on a new word from astronomy: *quasar*, one of the true mystery guests of the universe, whose nature and existence is still being explored and understood.

Part of the extraordinary discovery of that enormous supercluster that included at least eleven galaxies was that it also included at least eighteen quasars. First discovered in 1963 with the use of radio telescopes, quasars are thought to be the highly energized centers of very distant galaxies. The name originated as an acronym for "quasistellar radio source" but quasars are now described as "quasistellar objects" (QSO). Although the name means "starlike," because quasars appeared like stars, they are very different from stars. Extremely far away, and yet very bright, they emit more energy than one hundred giant galaxies. Quasars can shine with the brilliance of a trillion suns and, according to recent theories, are believed to be galaxies with very active and bright center objects, possibly powered by black holes. The farthest away are over 12 billion light-years from Earth. These mysterious powerhouses exist on the edge of the visible universe—at the beginning of time itself.

So, galaxies look like pinwheels and disks and squashed soccer balls. Why aren't space objects shaped like bananas or sugar cubes?

Newton's apple falling to the ground. The Moon falling toward Earth. Earth falling toward the Sun. The Sun falling toward the center of the Milky Way. The Milky Way falling toward Andromeda. The Local Cluster falling toward the Virgo Cluster.

One rule alone explains the behavior of the apple and the super cluster. Gravity. The Super Glue of the Universe. Gravitational attraction is one of the fundamental properties of matter, the force of attraction that acts between all objects because of their mass—that is, the amount of matter they are made of. Science now holds that there are four fundamental forces in the universe: the *strong* and *weak nuclear forces*, which operate inside atoms; the *electromagnetic force*, which gives matter structure; and *gravity*. Although it is the weakest of these four, gravity is the one most easily observed. Even a child can understand the basic idea of gravity when he falls off a swing at the playground. Gravity pulls objects that are on or near Earth back toward it, whether they are the Moon, a meteorite, or an arrow shot straight into the air. Gravity holds together the hot gases in the Sun. It keeps the planets in their orbits around the Sun, and it keeps all the stars in the galaxy in their orbits about its center.

The Earth is round because of gravity. The Sun, Moon, other planets, and stars are also round, because gravitational pull of every bit of matter on every other bit of matter pulls it into a ball—the shape that requires the least amount of energy to hold it together.

An explanation for gravitational force puzzled people for centuries. The ancient Greek philosopher Aristotle taught that heavy objects fall faster than light ones, and his view was accepted for centuries. But in the early 1600s, Galileo introduced a different concept. According to Galileo, all objects fall with the same acceleration (rate of velocity), unless air resistance or some other force slows them down. Galileo was onto something, but these motions were not correctly explained until Sir Isaac Newton showed that there is a connection between the force that attracts objects to Earth and the way the planets move. Newton realized that the same force that makes objects fall to the ground is what raises tides in the oceans, holds the Moon in orbit, and propels the planets on their endless trips around the Sun.

Newton based his work on the careful study of planetary motions made earlier by Tycho Brahe and Johannes Kepler. From laws discovered by Kepler, Newton showed how the Sun's gravitational attraction must decrease with distance. He assumed that the gravity of Earth behaves the same way.

Newton's theory of gravitation says that the gravitational force between two objects is proportional (or directly related) to the size of their masses. That is, the larger either mass is, the larger the force is between the two objects. The theory refers to mass rather than weight because the weight of an object on Earth is really the strength of the Earth's gravity on that object. On different planets, the same object has different weights, but its mass is always the same. The second part of the theory holds that the gravitational force is inversely proportional to the distance between the centers of gravity of the two objects squared (multiplied by itself). For example, if the distance between the two objects doubles, the force between them becomes a fourth of its original strength. To put it another way, this is why a comet speeds up as it gets closer to the Sun and slows down as it moves away from the Sun—the gravitational pull of the Sun on the comet weakens as it moves farther away.

In 1915, Albert Einstein threw a bit of a monkey wrench into the works by showing that Newton's theory did not hold for all things, like the orbit of Mercury around the Sun. In truth, the difference is very small and we can still rely upon Newton to explain most of the universe's behavior.

MILESTONES IN THE UNIVERSE
1895–1929

1895   James Keeler observes Saturn's rings and recognizes that they do not rotate as a unit, suggesting that they are formed of discrete particles.

Konstantin E. Tsiolkovsky, a Russian scientist, publishes his first scientific papers on space flight; he is referred to by the Russians as the "Father of Space Flight." In 1903, he will propose that liquid oxygen can be used for fueling rockets for space travel.

1896   The first lunar photographic album is published by the Lick observatory in California.

Samuel Pierpont Langley tests his steam-driven flying machine on the Potomac, flying for 1.2 kilometers (0.75 miles) before crashing.

1901   Annie Jump Cannon completes the Harvard Classification of stars.

The first electric typewriter, the first vacuum cleaner, the first safety razor are all produced; the first transatlantic telegraphic radio transmission is made.

1903 The first successful airplane is launched at Kitty Hawk, North Carolina by Wilbur and Orville Wright.

1905 Percival Lowell predicts the existence of a ninth planet with an orbit beyond Neptune.

Albert Einstein submits his first paper on the special theory of relativity, "On the Electrodynamics of Moving Bodies." His theory calls the speed of light constant for all conditions, and states that time passes at different rates for objects in constant relative motion.

On September 25, Einstein's second paper on relativity, "Does the Inertia of a Body Depend on Its Energy Content?" is published; it states the famous relationship between mass and energy: $E = mc^2$.

The first dial telephone is invented; the first airplane factory opens near Paris.

1906 A mysterious explosion occurs near Tunguska, Siberia, flattening a huge region and knocking down millions of trees; no meteorite remains are found in the area and the cause is never determined.

1909 Louis Blériot completes the first successful flight across the English Channel, flying from Calais to Dover in thirty-seven minutes.

1910 Eugene Ely becomes the first to take off in an airplane from the deck of a ship.

1911 A meteorite the size of a basketball kills a dog in Nakhla, Egypt, the only recorded case of a meteorite killing an animal; seventy-five years later scientists will decide that the meteorite originated on Mars.

1912 The *Titanic* sinks on its maiden voyage.

The ancient tradition of Chinese court astrologers is abolished.

1914 World War I begins.

Robert Goddard starts to develop experimental rockets.

1915   Albert Einstein completes his theory of gravity, known as the general theory of relativity.

German Fokker aircraft become the first airplanes to be equipped with a machine gun that can be fired between the blades of a moving propeller.

1918   World War I ends on November 11.

1919   British astronomers led by Sir Arthur Eddington observe a solar eclipse and announce that photographs of the eclipse confirm Einstein's theory of gravity.

"A Method of Reaching Extreme Altitudes," by Robert Goddard, is published by the Smithsonian Institute. In it, he suggests sending a small vehicle to the Moon using rockets. Goddard is ridiculed by the press.

1921   Einstein wins the Nobel Prize for Physics for his discovery of the law of photoelectric effect.

The word "robot" is coined by Czech playwright Karel Capek in his play about mechanical people, *RUR*.

1922   Aleksandr A. Friedmann predicts that the universe is expanding, basing his ideas on Einstein's theories.

1924   Edwin Hubble demonstrates that the galaxies are true independent systems rather than parts of the Milky Way system.

The Rocket into Interplanetary Space by Hermann Oberth is the first truly scientific account of space-research techniques; it introduces the notion of "escape velocity."

1926   Robert Goddard launches the first liquid-fuel-propelled rocket, which reaches a height of 56 meters (184 feet) and a speed of 60 miles (97 kilometers) per hour.

1927   Georges F. Lemâitre, a Belgian priest, proposes that the universe was created by the explosion of a concentration of matter and energy, which he called the "cosmic egg" or "primeval atom," the first version of the Big Bang theory of the origin of the universe.

Charles Lindbergh makes the first nonstop solo flight across the Atlantic in 33.5 hours.

The Society for Space Travel is founded in Germany; among its members are Wernher von Braun, who will develop the first rockets to travel in space.

1929 Edwin Hubble determines the distance of the Andromeda nebula from our solar system.

Hubble establishes that the more distant a galaxy is, the faster it is receding from Earth. "Hubble's Law" confirms that the universe is expanding.

Robert Goddard launches the first rocket carrying instruments; it carries a barometer, a thermometer, and a small camera.

What are stars and do they twinkle?

So galaxies are filled with stars, lots of them, more stars than most of us can really fathom. Does the number 10 billion trillion stars really mean anything to you? By one estimate, if everybody in the world divided up the stars equally and tried to count them one by one, each person would have to count more than one and a half trillion stars. At a rate of one thousand stars per second for twenty-four hours a day, it would take fifty years to count that many stars.

Here are some star basics:

- Most are unfathomably enormous, a word that doesn't do the stars justice. The Sun is a medium-sized star, but its diameter is more than one hundred times that of Earth. The largest stars have a diameter that is about one thousand times as large as the Sun's. However, there are also stars that are smaller than Earth, like neutron stars that have a diameter of 12 miles (20 kilometers).

- They're *really, really* far away. Even the largest stars look like pinpoints in the sky, because they are so distant from Earth. Besides the Sun, the nearest star is more than *25 million million* miles (40 million million kilometers) away.

- Stars appear to "twinkle" when we see them from Earth, because they travel through moving layers of air that surround the Earth. The Sun is a star and it doesn't twinkle. Stars would look like the Sun if we could see them all that close. The stars shine all the time, but we can see them only when the sky is dark and clear, that is, when the Sun that provides our light does not keep us from seeing them.

- At night, the stars appear to move across the sky—just as the Sun and Moon do. But the motion we perceive is actually caused by the rotation of Earth, not from the movement of the stars. The stars are moving, but their true motion cannot be seen because they ware so far from the Earth. It is comparable to looking at a jetliner high in the sky, which seems to move lazily through the sky, even though it may be flying at a thousand miles per hour while a slower, low-flying plane appears to move much faster. Astronomers measure the changing positions of stars, called *proper motion*, by comparing photographs taken at regular intervals.

- They are old, *really, really* old. Most stars began shining about 10 billion years ago, although new stars are still forming within the clouds of gas and dust of the Milky Way and other galaxies.

How do stars form and what makes them shine?

The life of most stars lasts billions of years. Obviously, no one has ever watched a particular star take shape, change, and, finally, die. However, astronomers have observed many different stars at various stages of their existence and developed theories of star formation based on known chemical and physical laws.

Let's go back to good old gas and dust. A star begins its life as a cloud of interstellar gas, which consists mostly of hydrogen mixed with dust. It is disturbed by some pressure and begins to collapse in on itself from gravitational forces. Such clouds can be seen as dark patches in front of the bright, distant stars of the Milky Way. The cloud may include the remains of a star that exploded, or it may be a collection of gases thrown from the surface of giant stars. Astronomers have never watched a new star flash into life, but several dark, ball-like interstellar clouds that may be new stars beginning to take shape have been observed.

Through millions of years, the cloud of gas and dust contracts as gravity pulls it together. As the material pulls together into a ball, the pressure of the gas increases and the particles are drawn in at an ever-increasing rate. As the collapsing cloud contracts, the gas at the center of the ball becomes extremely hot and the cloud has become what is called a *protostar*.

When the temperature at the center reaches extremes, the nuclear fusion reaction begins. Hydrogen atoms in the core begin to change into helium atoms, releasing great amounts of nuclear energy as they do. As a result of the reactions, a helium nucleus is created out of four hydrogen nuclei. When this nuclear fusion takes place, energy is released. A similar process occurs when a hydrogen bomb explodes. This energy heats the gas that surrounds the center. The gas begins to shine.

A star is born!

What is star dust?

"Earth to earth, ashes to ashes, and dust to dust" read the familiar burial words from the Book of Common Prayer. How poetic. How accurate. Space is not the Final Frontier after all, but the Great Recycling Bin.

Stars are born of the stuff of old stars. Some stars end their existence in explosions that create a cloud containing the helium and heavier elements that were in the former star. In time, the material in the cloud mixes with interstellar gas, and the enriched gas becomes the material out of which new stars are formed. For example, the Sun—and the Earth and other planets of the solar system—formed about 4.6 billion years ago out of material—the star dust—enriched by earlier generations of stars. Most of the dust in the clouds comes from the earlier generations of stars, which long ago completed their life cycles and blasted their material into interstellar space. The oxygen in the air, the iron in your blood, and the calcium in your bones—all the atoms in our bodies, except for the hydrogen ones—were formed in the interiors of stars that exploded long before the solar system formed.

Red giants, white dwarfs, and black holes—how do stars change and die?

It's a little like Hollywood. After a star begins to shine, it starts to change slowly, sort of like the charming and humble young actor whose ego takes over once he becomes a major star.

In space, after a star begins to shine, it also starts to change slowly. The speed of a star's change depends on how rapidly the nuclear energy-producing process takes place inside it. The speed of this process, in turn, depends on the mass of the star. The greater a star's mass, the higher its luminosity and temperature—and the faster it changes. The Hollywood analogy still seems to hold!

Stars that have a mass about ten times that of our Sun take a few million years to change. Smaller stars that have a mass about one-tenth that of the Sun take hundreds of billions of years to change. A star changes because its supply of hydrogen decreases. When this decrease occurs, the star's center contracts, and the temperature and pressure at the center rise. At the same time, the temperature of the outer part gradually drops. The star expands greatly and becomes a *red giant*.

What happens after a star's red-giant phase depends on how much mass the star contains. A star with about the same mass as the Sun throws off its outer layers, which can be seen as a glowing gas shell called a *planetary nebula*. The core that is left behind cools and becomes a *white dwarf*. A star with more than about three times the mass of the Sun becomes a *supergiant*. Elements as heavy as iron may be formed inside the star, which may then explode into a *supernova*. If less than three times the mass of the Sun remains after the supernova explosion, it becomes a *neutron star*. If more than three times the mass of the Sun remains, the star collapses and forms a *black hole*.

VOICES OF THE UNIVERSE:
DR. MICHAEL GARCIA, Harvard-Smithsonian Center for
Astrophysics (January 2001)

By detecting very little energy from these black hole candidates, we have new proof that event horizons exist. It's a bit
odd to say we've discovered something by seeing almost
nothing, but, in essence, this is what we have done.

Has anyone found a black hole yet?

A black hole has been called a black cat in a dark basement. A black
bird in the night sky. A vacuum cleaner, or a bathtub drain in space,
capable of sucking in everything around it. It has inspired movies,
myths, and head-scratching—an object in space whose gravity is so
great that, in theory, nothing can escape from it, including light. A
black hole, if there was sound in space, would probably make a very
big sucking sound.

Black holes are now thought to be the last evolutionary stage of
massive stars—at least ten to fifteen times as massive as the Sun.
Astronomers believe that a black hole forms when a massive star runs
out of nuclear fuel and is crushed under its own gravitational force.
While stars burn fuel, they produce an outward push that counters
the inward pull of gravity. But when the fuel is spent, the star can no
longer support its own weight, and collapses, exploding in a tremendous burst called a supernova. The exploding star throws off its outer
layers, and the remaining core continues to collapse to a small point
in which all of the star's mass is packed in with infinite density. The
gravitational force is so strong very near the black hole because all its
matter is located at a single point in its center. Physicists call this point
a *singularity*. It is believed to be far smaller in size than the nucleus of
an atom. Any light rays emitted by the star are wrapped around the
star. Because no light can escape, it is called a black hole.

There are thought to be two types of black holes—*non-rotating*
and *rotating*. Both have a spherical surface that is called an *event horizon*, from which nothing can escape. The radius of the event horizon
is used by astronomers to specify the size of a black hole. Once an

object passes through the event horizon, it disappears from our universe forever. Or, as Porky says, "That's all, folks."

Because black holes are invisible, astronomers have only indirect evidence of their existence. No one has discovered a black hole for certain, yet few doubt they exist. Most astronomers believe that the Milky Way, our galaxy, contains black holes. Most of their evidence comes from observations of X rays given off by a pair of stars that orbit each other, called *binary stars*. In recent years, using information from the Hubble Space Telescope and the orbiting Chandra X-ray Observatory, astronomers have found strong evidence for at least seven black holes in the Milky Way and in a nearby galaxy. Each system is a source of intense X-rays, proof that it contains either a neutron star, a very small star that has collapsed into a "superdense" state, or a black hole.

In 1994, astronomers used the Hubble Space Telescope to find evidence for a huge black hole in the center of the galaxy M87. This galaxy is in the constellation Virgo, and is about 50 million light-years from Earth.

More recently, NASA's two space observatories, Hubble and Chandra, provided what may be the best evidence yet for the existence of an event horizon, the defining feature of a black hole. An event horizon is the boundary around which nothing can escape. Black holes are the only objects that could conceivably have an event horizon. Studying X-ray data, separate teams of researchers announced early in 2001 new evidence for black holes. If an object were a black hole, a small amount of energy would escape before it actually reaches the event horizon. As one Chandra team member put it, "Seeing just this tiny amount of energy escape from the black hole source is like sitting upstream watching water seemingly disappear over the edge. The most straightforward explanation for our observations is that these objects have event horizons, and, therefore, are black holes."

Another researcher, Joseph Dolan of NASA's Goddard Space Flight Center, spent years in search of an event horizon and announced what he thought was evidence of one, also in early 2001. Examining data from an object in the Milky Way called Cygnus XR-1, Dolan saw bright flashes of ultraviolet sputter and disappear. In black-hole theory, as a blob of hot gas approaches the event horizon—the theoretical point of no return—the immense pull of gravity stretches the waves in the light so flat that they are no longer visible.

Theories, of course, can't be proven. They can only be disproved. No matter how many times the result of an experiment agrees with a theory, there is no certainty that the next experiment or evidence that comes along won't disprove it. But each time evidence is presented to agree with a theory, its validity survives and is strengthened. The recent evidence provides more support for the theory of black holes. As NASA's Dolan put it when he made his announcement, "If we were trying to convict Cygnus XR-1 of being a black hole in court, we'd win a civil case that only needs a preponderance of evidence, but not a criminal case that requires evidence beyond a reasonable doubt."

A GUIDE TO STAR TERMS

Stars are classified in several ways. For example, stars differ in brightness, color, size, and mass. Remember: Mass is the amount of material they contain, which is different from size or weight. A very large object can be mostly composed of gas, so its mass is small. A very small object can be composed of very dense matter and have much greater mass. For instance, a beach ball or a balloon can be very large but have a much smaller mass than a golf ball or a marble. (Also remember that mass and weight are not the same; the amount something weighs depends on its location in space; but mass remains constant no matter where the object is. You weigh more on Earth than you would on the Moon, but your mass is no different.)

This list of words and terms is a brief overview of some of the key concepts related to stars and their life cycles.

Binary Stars (also known as "binaries"): Pairs of stars moving in orbit around their common center of mass. Most stars are binary, or even multiple. For example, the nearest star system to the Sun, Alpha Centauri, is a binary star. There are several categories of binaries:

- **Eclipsing Binaries**—double stars, which consist of a pair of stars that move around each other. The stars move in such a way that one periodically blocks the other's light.

- **Visual Binaries**—stars that, when seen through a telescope, look like two stars revolving around each other. One revolution of these stars may take one hundred years.

- **Spectroscopic Binaries**—stars that look like single stars, even through a telescope. They are so close together that they cannot be seen separately, and are named for the spectroscope, an instrument that spreads starlight into a spectrum, the band of colors similar to a rainbow. Certain features of the spectrum identify the light as coming from a binary. Spectroscopic binaries complete their revolutions around each other in a few days or a few months.

Brown Dwarf: Essentially, a star that didn't make the grade, a brown dwarf is an object less massive than a star, but heavier than a planet. Difficult to detect, brown dwarfs do not have enough mass to ignite nuclear reactions at their centers. The first was spotted in 1995, in the constellation Lepus. It is twenty to forty times as massive as Jupiter but emits only one percent of the radiation of the smallest known star.

Main-Sequence Stars: "Ordinary" stars, like the Sun. They make up about 90 percent of the stars that can be seen from Earth. They include stars of all star-colors and many degrees of brightness. Main-sequence stars have medium-sized diameters. They are so much smaller than giants and supergiants that they are sometimes called main-sequence dwarfs. All main-sequence stars burn hydrogen into helium through nuclear fusion deep inside the star. Medium-sized stars, commonly called or pronounced "dwarf stars," are about as large as the Sun. Their diameters vary from about a tenth of that of the Sun to about ten times the Sun's diameter.

Nebula (plural, nebulae): A cloud of gas and dust in space, usually glowing due to reflected, or absorbed and radiated, light from nearby stars. These are considered the "factories" or birthplaces that produce new stars. Some nebulae (*planetary nebulae* and *supernova remnants*) are produced by gas thrown off from dying stars. Nebulae are classified according to whether they emit, reflect, or absorb light.

- **Emission Nebula**—a nebula, such as the Orion Nebula, that glows brightly because its gas is energized by the stars that have formed within it.

- **Reflection Nebula**—a nebula, such as the one surrounding stars of the Pleiades cluster, in which sunlight reflects off the grains of dust within it.

- **Dark Nebula**—a dense cloud, composed of molecular hydrogen, which partially or completely absorbs the light behind it, such as the Horsehead Nebula in Orion.

The Crab Nebula, the most famous supernova remnant named for its crablike shape, is a cloud of gas 6,000 light-years from Earth, in the constellation Taurus. It is the remains of a star that, according to Chinese records, exploded on July 4, 1054, as a brilliant point of light—now known to be a supernova. At its center is a pulsar that flashed thirty times a second.

Neutron Star: The next time you want to insult someone for being less than intelligent, you might try this: "You are as dense as a neutron star."

Neutron stars are the tiniest stars, a very small star that has collapsed into a "superdense" state in which the atoms have merged—their electrons have been squeezed into the protons, turning them into neutrons. (For those whose recollections of science class revolve around defective Bunsen burners and dissections of frogs, neutrons are one of the three chief components of an atom: protons and neutrons are contained in a central nucleus surrounded by electrons.) These small, superdense stars are thought to form when massive stars explode as supernovae, during which the protons and electrons of the star's atoms merge, owing to intense gravitational collapse. Neutron stars behave as normal stars do, but are so condensed that a fragment the size of a sugar cube would weigh as much as all the people on Earth put together. With as much mass as the Sun, they are so compact that they are only about 12 miles (20 kilometers) in diameter.

Nova (plural, Novae): This a somewhat confusing term at first, especially if you think "nova" means "new." If you think it's a kind of fish that goes on bagels, you have a lot of catching up to do. A nova is actually an existing star that brightens suddenly and very dramatically, remains bright for a few days, and then fades away, gradually returning to its dim appearance. Novae are thought to result from the explosion of matter that accumulates on the surface of a white dwarf (see below) in a binary system.

Planetary Nebula: Despite its name, it has nothing to do with planets. This is the ejected gaseous envelope of a red-giant star, one stage in the death of a star. This shell of gas is lit up by the ultraviolet photons that escape from the hot, white-dwarf star that remains. It is believed that this reject from a dying star becomes the material to form new stars, part of the great cosmic recycling bin.

Pulsar: Thought to be rapidly spinning neutron stars that emit regular pulses of energy, discovered in 1967.

Red Giant: A large, bright star, ten to one-hundred times the size of the Sun, with a cool surface, it is believed to be at the end stage of its life cycle. The relatively low surface temperature of this star produces its red color. As a star uses up its fuel, it expands, its surface cools, and it sheds its outer atmosphere (which becomes a planetary nebula). The diameter of Aldebaran, for example, measures about forty-five times that of the sun. More massive stars in their giant phase are referred to as *supergiants*.

Supergiant: The largest and most luminous type of star known, they are dying stars whose diameter is up to 1,000 times that of the Sun. A very massive type of star that has used up its hydrogen fuel and begun to expand and cool, they are the largest known stars, and include such stars as Antares and Betelgeuse. Antares has a diameter 330 times that of the Sun. Betelgeuse actually expands and shrinks. Its diameter varies from 375 to 595 times that of the Sun. Such supergiants as Betelgeuse shine with a low-temperature red color, though some supergiants, such as Deneb, shine with a blue light, indicating a higher temperature.

Supernova: An exploding supergiant, it is the explosive death of a star, which attains the temporary brightness of 100 million Suns or more. A supernova can shine as brightly as a small galaxy for a few days or weeks. Astronomers believe this may occur in a large galaxy about once every hundred years. The most famous supernova occurred in the Milky Way in 1054 and produced a huge cloud of rapidly expanding gas called the Crab Nebula. The Crab Nebula has a spinning neutron star—a pulsar—at its center.

Variable Star: These star's properties, such as brightness, vary at different times, usually due to pulsations in the star.

White Dwarf: A very dense, small, hot star in the last stage in the life of a star such as the Sun. After a red giant sheds its outer layers as a planetary nebula, this is the last stage of a dying star, creating small, superdense stars that are planet-sized, but as heavy as normal stars. White dwarfs make up 10 percent of the stars in the Milky Way. Most are hotter than the Sun but not very luminous, because they are so small.

Gravity has packed the electrons and protons in a white dwarf as tightly as physically possible; a spoonful would weigh several tons on Earth. Believed to be the shrunken remains of stars with exhausted energy supplies, they slowly cool and fade. Astronomers surmise that gravity within white dwarfs has shrunken them to their small size. The smallest white dwarf, van Maanen's Star, has a diameter of 5,200 miles (8,370 kilometers)—less than the distance across Asia. White dwarfs are much smaller than main-sequence stars, but their temperatures are higher than those of main sequence stars. White-dwarf stars no longer have a supply of energy from fusion. They will eventually fade away into cool dark embers, the most likely fate of our Sun in 5 to 7 billion years.

"Beetlejuice, Beetlejuice, Beetlejuice."
Who gets to name the stars?

That's "BET-el-jooze" to you, pal. Betelgeuse, thanks to the Michael
Keaton film *Beetlejuice*, is one of the few stars whose name anyone
recognizes, other than the North Star, whose name is really Polaris. It
is originally Arabic and means "armpit of the giant." Like many star
names, Betelgeuse is an Arabic word because after the collapse of
Rome, during the medieval period in Europe, most scientific astronomy was practiced by the Arabs. During the Renaissance, when the
great stores of Greek knowledge were reintroduced to the West, often
through Arabic sources, the Arabic names of stars were retained in
many cases. Even so, there were simply too many stars to name hap-
hazardly, and, in 1603, a systematic approach to identifying stars was
instituted by German astronomer Johann Bayer in his *Uranometria*.
Bayer assigned each star a lowercase Greek letter followed by the pos-
sessive name of the constellation in which the star was located. Since
then, as our grasp of the universe has expanded, so has our classification system. One of the great "stars" of astronomy, Annie Jump Cannon, a pioneering astronomer at the Harvard Observatory, once said,
"Classifying the stars is the greatest problem to be presented to the
human mind." Then she classified almost all 250,000 stars in the
Harvard College Observatory Star Catalog.

To finally coordinate and keep all the different classification systems straight, one group took over the job. In 1919, the International
Astronomical Union was founded to promote and safeguard astronomy. In 1930, the IAU became the recognized authority on classifying
and identifying all celestial objects—that includes everything else
astronomers turn up besides stars.

Only a few stars actually have names. Ancient stargazers named
the brightest stars, such as Betelgeuse and Rigel in the constellation
Orion. But astronomers and skywatchers rely upon different systems to
recognize the stars and other night objects. Most stars in a sky atlas are
referred to by their Bayer classification, in which a Greek letter
denotes the brightness of the star in a particular constellation ("alpha"
is brightest, "omega" is least bright) and there is a three-letter abbrevi-

ation for the constellation. In other words, Sirius, the brightest star in the night sky, which is in the constellation Canis Majoris, is known officially as αCma. Astronomers have found constellations with more more than twenty-four stars (the number of Greek letters), and have used numbers and letters from the Roman alphabet. As telescopes improved, new systems were devised to deal with all the space objects other than stars that were discovered. One of the later cataloging systems, the Messier Catalog, was compiled in the 1700s by French astronomer Charles Messier; Messier designations always begin with the letter M (Andromeda, for instance, is known as M31). In 1786, William Herschel published the *Catalogue of Nebulae*, which became the New General Catalog (NGC), devised to handle thousands of newly discovered deep-sky objects—star clusters, nebulae, galaxies. Its supplement is the Index Catalog (IC). While these listings are found in most skywatching books, all new designations for stars and other celestial objects are the province of the IAU—and they, apparently, can't be bought.

So every year, particularly around Valentine's Day, you'll hear advertisements for companies that allow you to name a star after someone as a gift. They promise to record the star name in some official record book and to send off a small plaque and certification documents. If you have purchased one of these gift stars, it may have been a much-appreciated gift but it has no standing in the world of astronomy. So, if you fell for that, watch out for the guys who want to sell you the Brooklyn Bridge or a piece of prime real estate in Florida. The best way to get something heavenly named for you or a loved one is to discover it—such as a new comet.

Are there other solar systems and planets out there?

Now that we have some handle on the vastness of the galaxies and the universe, it begs the obvious question. Are there other planets? Other Earths? Is it possible, as some scientists suggest, that Earth is an extraordinary accident, a one-in-many-billions shot? Or are there possibly other Earths out there in the incredible vastness we call the universe?

Until the early 1990s, the answer was "maybe." But an astonishing wave of discovery since 1995 has provided a more certain answer. There are certainly other planets out there. By early 2001, more than forty other "extrasolar planets" had been discovered, although some of these challenge the accepted idea of what a planet is, and none of them is remotely similar to Earth.

But that view is changing as well. A flurry of new evidence late in 2000 showed the likely existence of another nine or ten planets circling stars. As one researcher, Geoffrey Marcy of the University of California at Berkeley, put it at the time, "We're now at the stage where we are finding planets faster than we can investigate them and write up the results."

Some recent discoveries, reported in the January 4, 2001, issue of *Nature*, hold forth the possibility that several nearby stars may have some of the necessary building blocks of solar systems. Researchers have found unexpected amounts of hydrogen gas, critical to the formation of giant planets like Jupiter, circling three nearby stars. The study reported in *Nature* found enough hydrogen gas around one of the stars to form six Jupiters. Put this together with the idea that such gas giants are necessary for the formation of smaller Earth-like planets and you have the theoretical makings of planets forming solar systems. According to solar-system-formation theories, gas giants are one of those basic requirements, like the courses you had to take in school, for allowing the development of smaller planets in orbits that could potentially support life. The gas giants act as "sweepers," whose gravity cleans up potentially threatening asteroids and comets.

The three stars in the study are relatively young—between 8 and 30 million years old, compared with our Sun's nearly 5 billion years. Each of the stars is less than 260 light-years away.

Of course, the search for other planets and the growing numbers of planets being discovered raises The question that is probably most tantalizing to the average person who is less concerned with cosmology and the fate of the universe: Are we alone?

MILESTONES IN THE UNIVERSE
1930–1945

1930   Clyde Tombaugh discovers the planet Pluto on February 18.

1934   Engineer Wernher von Braun develops a liquid-fueled rocket that achieves a height of 1.5 miles (2.4 kilometers).

1936   Hitler's Germany builds a secret base for the construction of experimental liquid-fueled rockets, Peenemünde.

1937   Japan invades China.

The first rocket tests are performed at the Baltic research station at Peenemünde; one of the leaders of the project is Wernher von Braun.

Frank Whittel builds the first working jet engine.

The asteroid Hermes, probably a half-mile in diameter, passes a half-million miles from Earth—the closest approach of a large body (besides the Moon).

1938   Hans Bethe and Carl Friedrich von Weizsäcker independently propose a theory of the cause of energy produced by stars—nuclear fusion of hydrogen into helium; this basic concept is still accepted today with some modifications.

Otto Hahn splits the uranium atom, opening up the possibility of a chain reaction and atomic bombs; in March, Lise Meitner flees Nazi-controlled Austria for Sweden, taking the problem of uranium atom-splitting with her; a January 1939 paper she writes starts the drive to develop an atomic bomb.

1939   September 1: German and Soviet forces invade Poland, precipitating World War II. In August, Albert Einstein writes a letter to President Roosevelt that will lead to the U.S. effort to develop the atomic bomb (the Manhattan Project).

J. Robert Oppenheimer, who will be the lead scientist in the Manhattan Project, calculates that if the mass of a star is more than 3.2

times the mass of the Sun, a collapse of the star caused by the lack of internal radiation would lead to the star's mass being concentrated at a point, creating what would come to be known as a "black hole."

Pan American introduces regular commercial transatlantic flights.

1941   The Japanese attack Pearl Harbor on December 7, drawing the United States into World War II. One day earlier, President Roosevelt had signed the order leading to development of the bomb that would be dropped on Japan in 1945.

1944   The German armed forces begin to use the V-1 flying bomb, propelled by a jet engine, against the United Kingdom; in September, a liquid-fueled-rocket-propelled bomb, the V-2, goes into operation.

Carl Friedrich von Weizsäcker's theory of origins of the solar system is widely accepted; in it, small bodies called *planetesimals* are attracted to each other to form the planets.

1945   Germany surrenders on May 7; Hiroshima is bombed on August 6 with a uranium-based atomic bomb. A plutonium-based bomb destroys Nagasaki on August 9; Japan surrenders on August 14.

The White Sands proving ground for U.S. rocket research is established in New Mexico. It will soon be staffed by many former Nazi scientists and rocketeers who have been recruited by the U.S. military to help develop America's space program.

Robert Goddard dies.

PART IV

TO BOLDLY GO

O dark, dark, dark. They all go into the dark. The vacant
interstellar spaces, the vacant into the vacant.

T.S. ELIOT, *Four Quartets*, 1940

I am a passenger on the Spaceship Earth.

R. BUCKMINSTER FULLER
Operating Manual for the Spaceship Earth, 1969

The Truth is out there.

The X-Files

How did a basketball-sized satellite and a dog named Laika change history?

Was Wernher von Braun a war criminal?

Did men really walk on the Moon?

Was the film *Apollo 13* accurate?

Who were the "Mercury 13"?

What happened to the space shuttle *Challenger*?

Cell phones and Speedos: What have you done for me lately, NASA?

Back to the future: When John Glenn returned to space, was it good science or just a publicity stunt?

If you break a mirror in space, is it seven years of bad luck?

Space, toilets, and sex: How will the space station work?

Has anyone been abducted by aliens?

Who is looking for life?

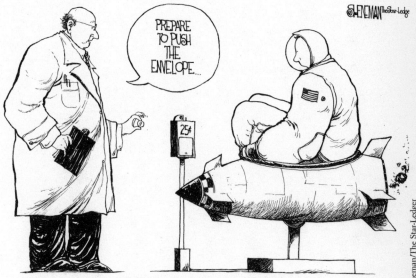

In April 1997, *Celestis I*, one of a new generation of privately launched commercial spacecraft, blasted off with an unusual cargo aboard. Instead of deploying communications equipment or weather monitoring instruments, this small satellite was carrying the cremated remains of twenty-four people. This mission was designed to carry out the first human burial in space. Among the remains consecrated to the depths of outer space were those of the *Star Trek* creator Gene Roddenberry, space-colonization advocate Gerard K. O'Neill, and sixties' psychedelic guru Timothy Leary. Ashes to ashes. Dust to space dust.

The universe, we now know, constantly recycles itself. It takes the leftovers from dead stars and turns them into new galaxies, stars, and planets. But is that all there is? Is space destined to become humanity's final landfill, the resting place for human remains, shipped off to space to relieve cemetery overcrowding on the home planet? Was Neil Armstrong's "one small step" really a giant leap for mankind—or morticians?

This chapter looks at a story that began when the first person gazed at the stars and conceived of going there someday. From the mythical Daedalus to Leonardo da Vinci and Kepler, to Jules Verne and H. G. Wells and the Wright Brothers, humans have dreamed of escaping Earth's hold. It took centuries, but we did finally move off the planet. What follows is the story of the practical pursuit of reaching for the heavens, an overview of the twentieth-century space programs that have allowed humanity to finally slip "the surly bonds of Earth," as poet John Gillespie Magee Jr. put it.

That pursuit of space, the "high frontier," has generally been accepted and cheered by the average earthling. But it is also a story filled with all the makings of an international conspiracy thriller—personal ambitions, political machinations, intrigue and espionage between nations, and many morally questionable decisions that never saw the light of day while the American media lauded the country's dazzling technical achievements.

As the space program moved forward, many questions were answered. But our missions to other worlds also raised a good many others. Perhaps the biggest question of all, one that has gotten more intriguing with several recent discoveries, is: Are we alone? As we plumb the seemingly immeasurable distances of spacetime, edging out from the safe shelter of Mother Earth, will we find other life?

VOICES OF THE UNIVERSE:
JOHN GILLESPIE MAGEE JR., *High Flight*, 1941

Oh! I have slipped the surly bonds of Earth
And danced the skies on laughter-silvered wings;
Sunward I've climbed, and joined the tumbling mirth
Of sun-split clouds,—and done a hundred things
You have not dreamed of—wheeled and soared and swung
High in the sunlit silence. Hov'ring there,
I've chased the shouting wind along, and flung
My eager craft through footless halls of air. . . .

Up, up the long, delirious, burning blue
I've topped the wind-swept heights with easy grace
Where never lark, or ever eagle flew—
And, while with silent, lifting mind I've trod
The high untrespassed sanctity of space,
Put out my hand, and touched the face of God.

John Gillespie Magee Jr., son of American missionaries in China, was attending school in England when the war broke out. He enlisted in the Royal Canadian Air Force and wrote this poem on the back of a letter to his parents back in the United States. While training, he died in air collision four days after the Japanese attacked Pearl Harbor. He was nineteen. The poem was made famous when President Reagan read it at a memorial service for the astronauts who died in the space shuttle *Challenger* disaster (see page 251).

How did a basketball-sized satellite and a dog named Laika change history?

Until October 4, 1957, it was still a dream. On that day, the Soviet Union launched *Sputnik*, the first artificial satellite to orbit Earth. It sent back a steady *beep-beep-beep* transmission that was picked up by radio listeners around the globe. The Soviets had stunned America and the world by becoming the first nation to successfully launch a man-made satellite into orbit around the Earth. It was a propaganda coup for the Soviet Union; their system, they proudly announced, had proved superior. As the official Soviet news agency, *Tass*, reported, "Artificial satellites will pave the way to interplanetary travel, and, apparently, our contemporaries will witness how the freed and conscientious labor of the people of the new socialist society makes the most daring dreams of mankind a reality." Score one for the Motherland.

The Western world reacted to the launch of *Sputnik* with surprise, shock, fear, and, at least among space scientists, a measure of respect. Soviet Premier Nikita S. Khrushchev ordered massive funding of follow-up projects that would continue to amaze and dazzle the world. In the United States, leaders vowed to do whatever was needed to catch up. The "space race" had begun.

As Tom Wolfe put it in *The Right Stuff*, his chronicle of the astronauts and the American space program,

"The 'space race' became a fateful test and presage of the entire Cold War conflict between the 'superpowers,' the Soviet Union and the United States. Surveys showed that people throughout the world looked upon the competition in launching space vehicles in that fashion . . . a preliminary contest proving final and irresistible power to destroy. . . . But in these neo-superstitious times it came to dramatize much more than that. It dramatized the entire technological and intellectual capability of the two nations and the strength of the national wills and spirits. Hence . . . John McCormack's rising in the House of Representatives to say that the United States faced 'national extinction' if she did not overtake the Soviet Union in the space race."

A month after *Sputnik*, the Soviets hit another one out of the park. A second satellite, *Sputnik 2*, carried a mixed-breed terrier named

Laika ("Barker") into space. In a pressurized compartment complete with food, water dispensers, and electrodes monitoring her vitals, Laika proved that animals could survive the unknown effects of microgravity and radiation. But in the rush to beat the Americans, the Soviets had not invested in the proper techniques for reentry and recovery. Laika was killed onboard *Sputnik 2* with an injection of poison, a sacrifice on the altar of science.

The success of the twin Soviet launches astonished America and set the country on a new path. The Soviet Union's success came during the height of Cold War paranoia, the "great fear," or "Red Scare," during which every move of the American government was predicated on the defeat of Soviet Communism. It was a time when paranoid or simply pernicious politicians, like Senator Joseph McCarthy, believed that Communists and Communism were everywhere, from Hollywood to American State Department to the Army. The Soviets had "the Bomb," a reality that many Americans attributed to "Red" spies who must have turned over stolen American secrets. There were plenty of Americans who believed that the Soviets must have stolen American secret satellite and rocket designs as well. Now the "Reds" had a way to drop atomic weapons on America's head. People of a certain age may not know what "duck and cover" meant to a generation of American schoolchildren who were drilled to hide beneath their school desks or descend to the school basement to face the wall in the event of a Soviet missile attack.*

But in the early days of this race, the American runner stumbled. Things went from frightening to disastrous. On December 6, 1957, a few months after the successful flights of *Sputnik* and Laika, America's answer to *Sputnik* blew up in its face on the launch pad. The liftoff of a *Vanguard* rocket was watched by millions in the country's first nationally televised countdown. *Vanguard's* maximum altitude: four feet.

---

*Including this author. What we didn't know, of course, was how futile these measures were. If you lived in the vicinity of New York City or any number of other major American cities, you were probably at ground zero for a Soviet missile, in which case "ducking" beneath a desk and "covering" your head with your arms was a ridiculous idea. At best, the desk would have been vaporized in an instant, along with you and your schoolmates. Or you'd be dead from the blast, heat, or radiation in a very short time.

Clearly, if the Soviets were so far ahead of America, there was something wrong with American education, morality, and determination. In response, the nation embarked on a massive program, bordering on the hysterical, to upgrade and modernize its schools, which, in the view of many politicians, had become soft and mushy. The post-*Sputnik* American crusade to pour money into its schools, particularly in the area of science and mathematics instruction, was meant to bridge the "learning gap" with the Soviet Union. In her book *Measuring the Universe,* historian Kitty Ferguson recalls that frenzy during her high school days: "The panic filtered down to the level of my high school in Texas which wasn't, evidently, as good at producing scientists as Soviet schools were. Better physics and math books and lab equipment were purchased, teachers went to workshops for retraining, and we were all urged to take more classes in these subjects. My younger brother's stock, as a fledgling physicist and computer whiz, went up. Mine as a classical musician went down. At the University of Texas, my later-to-be husband was called unpatriotic by his mathematics professor for choosing not to major in math" (*Measuring the Universe*).

Another key development was the decision that led, in 1958, to the creation of a civilian space agency called the National Aeronautics and Space Administration (NASA). NASA spearheaded America's space effort by absorbing various aviation researchers and military space laboratories. Although ostensibly a civilian organization, NASA was, from its beginnings, largely in the grips of military needs and desires and it would remain so; three quarters of the satellites launched by the United States during the Cold War decades of the sixties and seventies were military satellites.

The Soviet *Sputnik* victory and NASA's birth were the culmination of centuries of dreaming about the possibility of space flight. In the early 1600s, the German astronomer and mathematician Johannes Kepler (see part I), who had charted the path of the planets, became the first scientist to describe travel to other worlds in one of the earliest works of science fiction, *Somnium*. In this book, which was part scientific treatise, a man is lectured about the Moon by a demon conjured by his mother. (Recall that Kepler's own mother had been accused of witchcraft!) This demon is able to fly to the Moon during eclipses, and describes the Moon's appearance and snakelike inhabitants.

During the eighteenth and nineteenth centuries, serious writers began to propose fictional ideas for travel to other worlds. Among them was the "father of science fiction," an unsuccessful French stockbroker turned successful writer Jules Verne (1828–1905). His 1865 novel *From the Earth to the Moon* not only inspired an early generation of rocketeers but was eerily prescient in some of its predictions, including the location of a space launch in Florida, a rocket fired by a giant cannon, and a "splashdown" in the Pacific Ocean. Among Verne's readers were some scientists who began to take the idea seriously in the twentieth century. In 1903, Konstantin E. Tsiolkovsky, a poor Russian high school teacher, deaf in one ear, completed the first scientific paper on the use of rockets for space travel. Unlike the previous century's fictions and fantasies, Tsiolkovsky's paper was grounded in hard science and proved that space travel was achievable. When Stalin came to power, he put the obscure schoolteacher in charge of a new Soviet program that was supposed to transform nature by constructing the Soviet utopia. (This effort was undertaken with extensive damage to the Soviet environment and at tremendous cost to the Soviet people.) Several years later, American rocket pioneer Robert H. Goddard and Transylvanian-born scientist Hermann Oberth were pursuing their interest in space travel with complete seriousness and dedication. Working independently, these three men addressed the technical problems of rocketry and space travel, and they are collectively known as the fathers of space flight.

The greatest pioneer of American rocket science, Robert Goddard (1882–1945) was born in Worcester, Massachusetts, and became fascinated with space flight, as so many other rocketeers did, when he read *War of the Worlds* as a boy in 1898, as it was serialized in a Boston newspaper. In 1919, Goddard, then a professor at Clark University in Massachusetts and a devoted apostle of space travel, explained how rockets could be used to explore the upper atmosphere, in a paper for the Smithsonian Institute called, "A Method of Reaching Extreme Altitudes." The paper also described a way of firing a rocket to the Moon. Goddard's ideas were met with a wave of ridicule and dismissal in the popular press. Goddard more or less withdrew from public view, but he also found a willing listener in Charles Lindbergh, the greatest hero of American flight, who arranged for some private financing. With it, Goddard built and launched the world's first liquid propellant rocket on his aunt Effie's farm in Auburn, Massachusetts, in 1926. By 1935, working by himself, Goddard was building and flying liquid-fueled rockets near Roswell, New Mexico.

Around the same time, in a 1923 book called *The Rocket into Interplanetary Space*, Hermann Oberth discussed the technical problems of space flight. It quickly became a widely read classic in Europe, although Goddard believed it had been taken from his own work. But Oberth had been developing his ideas independently, and one part of the book described a space ship, space travel, and a crude space station. His work inspired a Society for Space Travel, founded in Berlin in 1927. In 1929, Oberth worked as a technical advisor to the famed German film director Fritz Langon on a movie titled *Girl to the Moon*. Around this time, he also met a young German rocket scientist whose dreams of space flight had been inspired by reading Oberth's books. His name was Wernher von Braun. And both men would help Germany build rockets during World War II.

VOICES OF THE UNIVERSE:
TOM LEHRER, political satirist of the 1960s

Don't say that he's hypocritical
Say rather that he's apolitical,
"Once the rockets are up, who cares where they come
    down?
That's not my department," says Wernher von Braun.

Some have harsh words for this man of renown,
But some think our attitude should be one of gratitude,
Like the widows and cripples in old London town
Who owe their large pensions to Wernher von Braun.

Was Wernher von Braun a war criminal?

Throughout history, technology has been the bedfellow of weaponry. In ancient Greece, Archimedes supposedly invented a great reflecting shield that could set enemy ships on fire. Galileo designed a device to improve the accuracy of cannons. And twelfth-century Chinese inventors devised "fire arrows" with gunpowder to be used against their enemies. Although the image many people hold of "scientific research" is one of earnest, white-jacketed scientists bent over microscopes in hopes of improving the lot of humanity, the picture doesn't fit with history. The largest scientific project ever mounted was the "Manhattan Project," the American effort during World War II to build and perfect the atomic bomb—and to do so before the Germans, Japanese, or the Soviets, then America's wartime allies, did. Employing thousands of workers around the country, the Manhattan Project built whole cities such as Oak Ridge, Tennessee, and Hanford, Washington, as well as transformed an out-of-the-way boys' school in Los Alamos, New Mexico, into the maze of secret labs where the atomic bomb was born. Throughout history, and especially during the twentieth century, the military has always been, as famous bank robber Willie Sutton once said, "where the money was"—for biologists who developed germ weapons, chemists who developed chemical weapons, or physicists who developed nuclear weapons. And to many a researcher who dreamed of reaching the sky, the space program was no different.

In a small irony of history, the Germans began to develop a rocketry program because the Treaty of Versailles, which ended World War I in 1918, limited the size of the German army and its armaments, including production of artillery. Since rockets did not count as artillery, research in rocketry was funded by the German military in the years before World War II. During World War II, German rocket experts developed the A-series and V-2 guided missiles. Thousands of these "vengeance weapons" were launched at European cities such as Paris and Antwerp, but especially London, causing widespread destruction and killing more than twelve thousand people, most of them civilians. The Nazi rockets were the product of a program led by Wernher von Braun.

The aristocratic son of a wealthy German gentleman farmer who became Agriculture Minister in the Weimar Republic, Wernher von Braun (1912–1977) was born in what is now Poland. Fascinated with astronomy from childhood, he was a college student when the German army offered to subsidize his college education, and he became technical director of the German rocket program in 1932. In 1933, the Nazi Party took control of Germany. Taking note of von Braun's early successes, the German army and air force, or Luftwaffe, committed to fund rocket development, and a base called Peenemünde was built at a remote site on the Baltic Sea. Ordered to join the Nazi party in 1937, von Braun did so, just as he would later join the SS, Hitler's dreaded Nazi political enforcers, on the orders of Heinrich Himmler. In attempting to justify von Braun's actions, the kindest explanation is to say that ignoring an order from Himmler in those days was akin to committing suicide—not career suicide, but actual suicide. In fact, in 1944, von Braun and several members of his staff, including his younger brother Magnus, were arrested by the SS and accused of treason. They were held for two weeks until their release.

After Germany invaded Poland in 1939, and World War II began, von Braun's efforts focused on the A-2 rocket. British intelligence learned of these plans and, in 1943, bombed Peenemünde. Production was then moved to a secret site called Mittelwerk, also known as Dora. Under von Braun, Nazi Germany developed the A-4 rocket into what became known as the V-2 (or Vengeance Weapon 2), first used against London on September 8, 1944. It was built with slave labor at Dora. The horrific conditions of the camp rivaled the worst of the more renowned concentration camps such as Auschwitz and Treblinka. Half of the estimated sixty thousand Russian, Polish, French, and Jewish prisoners forced to work at Dora died at the facility, many of rampant diseases, many others by execution. That was far more than the number killed by the rockets themselves. The Nazi rockets were expensive and, ultimately, did little to change the war effort. Some military historians and Nazis, including Albert Speer, "Hitler's Architect," later made the case that the money that went to the rocket program might have been spent far more effectively elsewhere by the German High Command, especially to fend off the allied bombing of Germany that helped turn the tide of war.

Facing defeat as the war against Germany was winding down in 1945, Wernher von Braun and other German rocket engineers and scientists realized they had to make a choice: struggle in the ruins of the Third Reich, where rocketry and other scientific programs would be unthinkable, or play "Let's Make a Deal" with the Soviets or Americans. On May 2, 1945, the day after Hitler committed suicide, von Braun and other team members surrendered to the American army. Many of these scientists, eager to continue their research on rocketry and finding enthusiastic support in the American military, went to work for the U.S. government to help develop missiles. Without much debate and, in the military equivalent of "under the cover of darkness," von Braun and several hundred other German rocket scientists who had built weapons for Hitler with slave labor were secretly slipped into the United States in what was called "Operation Paperclip." The American army had secured materials for V-2s, which were also brought to the United States. As Dennis Piskiewicz writes in *The Man Who Sold the Moon*, a damning account of von Braun's wartime career, "With the rocket parts, the documentation, and the rocket scientists, the United States Army had the beginnings of its own rocket program."

Hundreds of other German scientists, researchers, and engineers went to work on other military projects, such as the jet plane and advanced torpedoes. Initially brought to America to aid the war effort against Japan, which continued until the atomic bombs were dropped on Hiroshima and Nagasaki in August 1945, many of them found a home in the United States and stayed on as a new war began—a "Cold War" with the Soviet Union, America's wartime ally against Hitler. There was little moral hand-wringing over the dilemma of hiring the enemy. It was either get them to work for the United States, or they would work for the "godless Communists" of the Soviet Union who had scooped up their fair share of German scientists in the war's closing days. Those former Third Reich rocketeers would help the Soviets stun the world with *Sputnik* in 1957, a little more than a decade after the war was over.

In *The New Ocean*, a dramatic, definitive, and all-encompassing history of the age of space flight, William E. Burrows devastatingly sums up the Faustian bargain made by the German scientists—and, by extension, America's military and political leadership:

"The evidence indicates that most of the German rocketeers, like their Soviet counterparts who joined the Communist Party, were ideologically apathetic. Their party cards were tickets to work and to social acceptance in an otherwise severely restrictive society. The cards put them safely on the inside in a place and a time when being on the outside could be abidingly dangerous. *It was easier to look at their drawing boards than inside their souls*" (emphasis added).

Over the course of the next thirty years, von Braun became the handsome, smiling poster boy for the American space program. One of its most enthusiastic cheerleaders and greatest salesmen, he wrote inspiring articles about space for magazines like *Collier's*, had a regular stint on the Walt Disney show to talk about the future of space, and became a technical advisor to Disneyland's Tomorrowland. More significantly, he helped design and build America's first long-range ballistic missile in 1953 and put the first U.S. satellite into space in 1958. In 1960, he joined NASA and spent most of the next decade designing the *Saturn 5* rocket that eventually took *Apollo's* astronauts to the Moon. In 1964, Stanley Kubrick's film *Dr. Strangelove* opened, featuring a brilliant but unsettling scientist who addressed the president as "Mein Füehrer," clearly modeled on von Braun.

Wernher von Braun's life as a Nazi was conveniently glossed over as his contributions to America's space program grew. Eventually, he was lionized by supporters of the space race, in and outside the government. Neither he nor most of the other German rocketeers who came to America ever had to answer publicly for their past. With one exception. Just as the American government seemed to have no qualms about hiring the former Nazi rocket designers, the United States later proved unwilling to stand by the decision. Arthur Rudolph, who had supervised the operations at Dora, was among the chief designers of the *Saturn 5* rocket, which took America's *Apollo* astronauts to the Moon. In 1982, Rudolph was identified by the U.S. Justice Department's Office of Special Investigations (OSI), formed after New York Congresswoman Elizabeth Holtzman prodded the justice department to track down Nazi war criminals living in the United States. Rather than attempting to defend Rudolph, the U.S. Government offered him a deal: to face a trial and risk the loss of a NASA pension, or renounce his citizenship. To avoid prosecution, Rudolph agreed to leave his home in San Jose and, in 1984, at age seventy-seven, went into self-exile in Germany while trying to regain his American citizenship. He died in 1995, having gone, in the space of fifty years, from a Nazi to an unsung hero of the American space program and back to a Nazi war criminal.

"The OSI investigated other living members of the German rocket team, but took no action against them. It did not investigate von Braun with the same vigor it pursued Rudolph or the others," according to historian Dennis Piskiewicz. But Piskiewicz also leaves no doubt as to his view of von Braun's culpability. He wrote, "Wernher von Braun willingly joined the Nazi cause and was an accomplice in its crimes in order to build rockets and to promote his own career. . . . Because of his tireless promotion, he was the man who sold the Moon. Sadly, because of his complicity with the Nazi cause, he also sold his soul to reach that goal" (*Wernher von Braun: The Man Who Sold the Moon*).

MILESTONES IN THE UNIVERSE
1946–1969

1946   The first meeting of the United Nations takes place.

The first all-purpose, all electronic computer—known as ENIAC—is completed.

1948   The Hale Telescope is completed at Palomar, California.

Austrians Herman Bondi and Thomas Gold, and British astronomer Fred Hoyle advance the steady-state theory of the universe, the chief rival to the Big Bang theory. In it, the continual creation of matter powers the universe.

1949   A rocket-testing ground is established at Cape Canaveral, Florida.

The first rocket with more than one stage is created when a small rocket is added to the top of one of the German V-2 rockets captured after World War II; launched from White Sands, New Mexico, it reaches a height of 240 miles (400 kilometers).

1950   Troops from North Korea invade South Korea; the United Nations intervenes in the Korean War.

Jan Hendrik Oort proposes that a great cloud of material orbiting the Sun far beyond the orbit of Pluto is the cause of comets; material from the Oort Cloud, as it is known, becomes dislodged from time to time and falls toward the Sun.

Commercial color television begins in the United States.

1955   Albert Einstein dies in Princeton, New Jersey on April 18.

1957   *Sputnik*, the first artificial Earth satellite, is launched by the Soviet Union on October 4.

In November, the Soviet Union launches *Sputnik 2*, a second satellite; this one contains a live dog, Laika.

1958   In January, the first U.S. satellite, *Explorer 1*, is launched by Wernher von Braun's team in America's first successful orbit around the Earth.

Eugene Parker demonstrates that "solar wind," particles thrown out by the Sun, causes the tail of a comet to point away from the Sun.

The orbit of a *Vanguard* satellite is used by Aloysius O'Keefe to show that the Earth is slightly pear-shaped, with a bulge of about 50 feet (15 meters) in the southern hemisphere.

1959   A series of Russian satellites aimed at the Moon meet with varied success. One goes into orbit around the Sun; a second crashes on the Moon's surface; a third, *Luna 3*, reaches the Moon and sends back the first photographs of the far side of the Moon in October.

In April, the United States launches its first military spy satellite, *Discover 2*.

1960   The first American weather satellite and first navigation satellites are launched.

1961   Soviet cosmonaut Yuri Gagarin becomes the first man in space, when he orbits the Earth on April 12 aboard *Vostok I*. Gagarin lands separately from the spacecraft after ejecting at 1 hour and 48 minutes, a procedure followed by all subsequent *Vostok* pilots.

Alan B. Shepard Jr. becomes the first American astronaut in space as his Mercury 3 capsule *Freedom* 7 completes a fifteen-minute suborbital flight on May 5.

Virgil I. Grissom flies his *Liberty Bell* 7 capsule on a sixteen-minute suborbital flight to become the second American in space on July 21.

25-year-old Soviet cosmonaut Gherman Titov, the youngest person in space, orbits the Earth seventeen times in 25 hours and 18 minutes on August 6, becoming the first man to sleep in space.

1962   On February 20, John Glenn Jr. becomes the first American to orbit the Earth in his capsule, *Friendship* 7.

M. Scott Carpenter completes three orbits of Earth in his capsule *Aurora* 7 on June 24.

Telstar is launched (July 10); the first active communications satellite, it relays the first transatlantic television pictures.

The Cuban Missile Crisis takes place as the United States blockades Cuba to prevent Soviet-guided missiles from being deployed on the island.

The U.S. space probe *Mariner* 2 passes within 22,000 miles (35,500 kilometers) of Venus (December); it is the first man-made object to voyage to another planet. On January 3, 1963, contact with the probe is lost at 54 million miles (87 million kilometers) from Earth.

Walter M. Schirra orbits the Earth six times in *Sigma* 7 on October 3.

1963   *Syncom* 2, launched on February 14, becomes the first geosynchronous satellite; its orbit matches the Earth's rotation so that it stays directly above one location on Earth's surface.

L. Gordon Cooper completes twenty-two orbits of Earth in his thirty-four hour flight aboard *Faith* 7 (May 15).

Two Soviet spacecraft with people aboard are placed in orbit simultaneously.

Soviet astronaut Valentina Tereshkova becomes the first woman in space, making thirty-eight orbits in seventy-eight hours (June 16).

On November 22, President John F. Kennedy is assassinated.

1964 U.S. space probe *Ranger* 7 takes the first good close-ups of the Moon; 4,316 pictures are relayed back to Earth.

1965 Arno Penzias and Robert W. Wilson accidentally find the radio wave remnants of the Big Bang while trying to refine their radio equipment; their discovery provides convincing support for the Big Bang theory, which is currently supported by most astronomers.

Soviet cosmonaut Aleksei A. Leonov takes the first "space walk," a trip outside his spaceship while wearing a spacesuit.

Virgil I. (Gus) Grissom—the first man to return to space—and John W. Young, America's first two-man crew, orbit the Earth three times (March 23).

(June) James A. McDivitt and Edward H. White orbit the Earth six times in *Gemini* 4 spacecraft; White makes the first American space walk.

Mariner 4 reaches the neighborhood of Mars, passing within 7,500 miles (12,000 kilometers) of the planet and producing the first images of the planet.

L. Gordon Cooper and Charles Conrad begin a 190-hour, 120-orbit mission on August 21; its purpose is to test the feasibility of a lunar mission.

Frank Borman and James A. Lovell Jr. are launched (December 4) for a thirteen-day mission during which they will make the first space rendezvous, with Walter M. Schirra and Thomas Stafford who are launched on December 15.

1966   The Soviet space probe *Luna* 9 lands on the Ocean of Storms (February 3), the first "soft landing" on the Moon.

The U.S. Environmental Science Services Administration launches the first weather satellite capable of viewing the entire Earth.

The Soviet space probe *Venera III* becomes the first man-made object to land on another planet when it reaches Venus (March 1).

Soviet spacecraft *Luna 10* orbits the Moon.

Surveyor I soft-lands on the Moon (June 1) and returns pictures of the lunar surface.

1967   On January 27, American astronauts Virgil I. (Gus) Grissom, Edward H. White, and Roger B. Chaffee die in a fire during a ground test of an *Apollo* spacecraft.

Soviet cosmonaut Vladimir M. Komarov dies (April 24) when his *Soyuz I*'s parachute lines get tangled and the spacecraft crashes.

Surveyor 3 scoops and tests lunar soil.

1968   *Surveyor 7* becomes the first space vehicle to land undamaged on the Moon (January 9). It sends back 21,000 photos of the lunar surface.

The Soviet *Zond 5* spacecraft becomes the the first man-made object to travel around the Moon and back to Earth.

Soviet cosmonaut Yuri Gagarin—the first man in space—dies (March 27) in the crash of a test plane.

Walter M. Schirra, Donn F. Eisle, and Walter Cunningham begin the first three-man Apollo mission (October 11).

On December 21, Frank Borman, James A. Lovell Jr., and William Anders become the first man-manned crew to orbit the Moon in *Apollo* 8.

1969   In the first space link-up of two manned vehicles, the Soviet *Soyuz 4* and *Soyuz 5* dock and transfer crews (January 14).

An *Apollo* 9 mission launched (March 3) with astronauts James A. McDivitt, David Scott, and Russell Schweickert testing all the lunar hardware for the planned lunar landing.

On May 18, the *Apollo 10* crew of Tom Stafford, John Young, and Eugene Cernan fly their lunar module to within 9 miles (14.5 kilometers) of the lunar surface.

Apollo 11 lands on the Moon: On July 20, American astronaut Neil Armstrong becomes the first human being to stand on the Moon; crewmate Buzz Aldrin is right behind him, while Michael Collins orbits above.

A second lunar mission, *Apollo 12*, is launched (November 14). More than fifteen hours are spent exploring the lunar surface.

VOICES OF THE UNIVERSE:
President JOHN F. KENNEDY,
Address to a Joint Session of Congress, May 25, 1961

I believe this nation should commit itself to achieving the goal, before this decade is out, of landing a man on the Moon and returning him safely to Earth.

While the Apollo Moon program and the successful Moon landing in 1969 became the crown jewel of the American space program, it wasn't always popular. A Gallup poll taken immediately after Kennedy's speech showed that 58 percent of Americans opposed the idea. And many in the public and politics opposed the expenditures going to space exploration, a debate that continues today.

Did men really walk on the Moon?

That question is not as silly as it might sound. There are people who still believe that the entire Apollo program and the Moon landings of the early 1970s were staged. This old conspiracy rumor got fresh life when Fox television, the people who brought you *The X-Files* and

Alien Autopsy, aired a program called *Conspiracy Theory: Did We Land on the Moon?* In the frenzied quest to beat the Soviets, the show contended, the entire Moon program was faked. One of the most commonly cited pieces of "evidence" for this belief—there are no stars in the pictures of astronauts standing on the Moon. Well, guess what? There are no stars in the pictures of Earth in the daytime. The Sun is shining too brightly for us to see the stars, and it is no different on the Moon, where sunshine and camera backlighting obscured any stars from the photos.

Other evidence? Moon rocks are just ordinary rocks. Well, they are not. Moon rocks are completely unique and lack many of the common substances found in Earth rocks.

So the easy answer is "Yes, men went to the Moon," unless thousands of people and a dozen astronauts have all conspired in a lie of cosmic proportions.

How, why, and whether it was a good idea is a different set of questions.

When the Soviet Union launched the world's first man-made satellite, *Sputnik*, the world was changed. During the early years of the space-age contest between the Soviet Union and the United States, success in space became one more measure of both countries' positions in science, engineering, and national defense. The posturing, in retrospect, was a little like two neighbors trying to upstage each other's Christmas-light displays, or two adolescent boys trying to impress the same girl. Except that those analogies don't properly contain the danger this competition posed to the entire world. These were adolescent boys who could literally end the human race in an instant of what was then called MAD, Mutually Assured Destruction.

As the intense Cold War rivalry passed from one decade to the next, careening dangerously into surrogate "hot wars" around the globe, the competition spilled into every aspect of geopolitics, international culture, and even the Olympics, as well as the "Space Race." It was displayed in dangerous showdowns such as the Cuban Missile Crisis (depicted in the recent film *Thirteen Days*) or in the defection of the Bolshoi ballet dancers. Throughout the 1960s and 1970s, this one-upsmanship between the superpowers, which often emphasized showmanship over science, still drove both nations to enormous achievements and sometimes dubious pronouncements. When President Kennedy made his famed speech promising a man on the Moon, there was plenty of amused and dismissive reaction, including the predictable refrain that America had more pressing needs at home than sending men to the Moon. The voices of disinterest and dissent ranged from artist Pablo Picasso, who said, "It means nothing to me," to journalists and social critics such as Amitai Etzioni who belittled the space program in a book called *The Moon-Doggle*.

The actual work of getting there had preceded Kennedy's election in 1960. Part of the mission included the ill-informed notion held by some of America's military men that the Moon would provide a base from which to launch missiles aimed at the Soviet enemy. Although it seems almost laughably impractical, this idea suggests the level of paranoia that pervaded American defense policies—and space policies—at the end of the 1950s. And by the end of that decade, both the United States and the Soviet Union began to launch probes toward the Moon. The first probe to come close to the Moon was *Luna 1*. Launched by the Soviet Union on January 2, 1959, it passed within about 3,700 miles (6,000 kilometers) of the Moon. The United States conducted its own lunar fly-by two months later with the probe *Pioneer 4*. The Soviet *Luna 2* probe, launched on September 12, 1959, was the first probe to hit the Moon, and, a month later, *Luna 3* circled behind the Moon and photographed its hidden far side.

Then, on April 12, 1961, the Soviets took another giant step ahead of the Americans. They put the first human in space, a twenty-seven-year-old test pilot named Yuri Gagarin who had had once planned to study tractor construction in a Moscow steel plant. Launched aboard a craft later referred to as *Vostok 1*, Gagarin orbited the Earth once in 108 minutes and returned safely. An autopilot device controlled the spacecraft during the entire flight. A twenty-five hour, seventeen-orbit flight by cosmonaut Gherman Titov aboard *Vostok 2* followed in August of that year.

Racing to keep pace, the American Mercury program made its first manned flight on May 5, 1961, when a Redstone rocket launched astronaut Alan B. Shepard Jr. in a capsule named *Freedom 7*. Shepard flew a fifteen-minute suborbital mission. It was followed by a second suborbital flight on July 21, 1961, by astronaut Virgil I. Grissom. This mission almost ended tragically—and still engenders controversy. After splashing down in the Atlantic Ocean, Grissom's capsule's side hatch opened too soon and the spacecraft rapidly filled with water.

Grissom managed to swim to safety but there were questions as to whether Grissom had caused the problem by blowing the hatch too soon. The unlucky Grissom was one of three astronauts who became America's first astronaut casualties when he and two other veterans of the Mercury program, Roger Chaffee and Edward White, died a fiery death on January 27, 1967, preparing for the first test mission of the Apollo craft that would take Americans to the Moon. An electrical short circuit probably started a fire in their capsule, where the pure oxygen atmosphere caused it to burn fiercely. Although NASA told the press that the three men had died instantly, the grim scene inside the capsule told a different story. The remains of two of the astronauts who were desperately trying to open a hatch were difficult to separate from each other. A few months after this tragedy, the Soviet space program also suffered a disaster when a *Soyuz 1* capsule was launched with Vladimir Komarov aboard as pilot. It was supposed to link up with a second manned spaceship, but *Soyuz 1* developed problems, and a parachute failure caused the capsule to crash, killing Komarov.

Playing catch-up with the Soviets, the American program struck gold on February 20, 1962, when John H. Glenn Jr. became the first American to orbit the Earth. Glenn completed three orbits in less than five hours. Perhaps the most cherished of the American Jedi Knights known as astronauts, Glenn became America's hero. After his space career ended, he entered politics, became senator from Ohio, attempted a run for the presidency, and then, in a mission that generated as much skepticism and derision as it did admiration and public fascination, returned to space, at age seventy-seven, aboard the space shuttle in 1998.

By 1963, the Soviet Union was testing lunar hard-landers, succeeding after a string of many failures, with *Luna 9* in January 1966. The U.S. Surveyor program made a series of successful soft landings beginning in 1966, and followed with five probes called Lunar Orbiters to photograph the Moon's surface.

After considering several proposals for a manned lunar mission, NASA settled on a plan known as Lunar-Orbit Rendezvous. A spacecraft consisting of three parts—a command module (CM), a service module (SM), and a lunar module (LM)—would carry three astronauts to an orbit around the Moon. One astronaut remained in the cone-shaped command module, which would be the spacecraft's main control center. The service module would contain fuel, oxygen, water, and the spacecraft's electric power system and propulsion system. These two units would be joined for almost the entire mission as the command/service module (CSM), orbiting the Moon, while two astronauts descended to the lunar surface in the lunar module—consisting of a descent stage and an ascent stage. Both stages would descend to the lunar surface as a single unit, but only the ascent stage would leave the Moon.

A *Saturn 5* booster would launch the spacecraft toward the Moon where it would go into a lunar orbit. The lunar module would separate and carry the two astronauts to the surface. By late 1968, the United States had redesigned the Apollo command-service module and it was ready to go, but the lunar module remained far behind schedule. Learning about Soviet preparations for a manned lunar fly-by, NASA decided to beat the Soviets to the punch with a manned mission to orbit the Moon, without a lunar module. The orbital mission would also test navigation and communication around the Moon.

Apollo 8, the first manned expedition to the Moon, blasted off from the Kennedy Space Center near Cape Canaveral, Florida, on December 21, 1968, carrying astronauts Frank Borman, James Lovell, and William A. Anders in what one NASA official called, "The single greatest gamble in space flight then, and since." Part of the mission involved twenty minutes spent on the far side of the Moon, out of radio transmission. It was the longest twenty minutes many in the space program had ever experienced, and it was finally over when radio contact was reestablished. It was Christmas Eve, and the three astronauts, in a deliberate tweak at the "godless" Communists, read the Creation account from the Book of Genesis. With a burn of the engines, which took them out of lunar orbit, Lovell reported, "Please be informed there is a Santa Claus. The burn was good." They landed in the ocean on December 27, 1968.

On July 16, 1969, *Apollo 11* blasted off carrying three astronauts—

Neil A. Armstrong, Edwin E. Aldrin Jr., and Michael Collins. It was the first mission to land astronauts on the Moon. The first two stages of a *Saturn 5* rocket carried the spacecraft to an altitude of 115 miles (185 kilometers) and a speed of 15,400 miles (24,800 kilometers) per hour, just short of orbital velocity. The third stage fired briefly to accelerate the vehicle to the required speed. It then shut down while the vehicle coasted in orbit. Then the astronauts began an extraordinary series of maneuvers. They checked the spacecraft and lined up the flight path for the trip to the Moon. The third stage of the rocket was restarted, increasing the speed to 24,300 miles (39,100 kilometers) per hour, the escape velocity to leave Earth's gravity for the Moon. On the way to the Moon, the crew pulled the command-service module away from the *Saturn* rocket, turned the module around and docked it to the lunar module, which was still attached to the *Saturn*. Then they pulled free of the *Saturn*.

For three days, *Apollo 11* coasted toward the moon. As the spaceship traveled farther from Earth, the pull of the Earth's gravity became weaker and the craft slowed down. By the time the ship was 215,000 miles (346,000 kilometers) from Earth, its speed had dropped to 2,000 miles (3,200 kilometers) per hour. Then lunar gravity took over and the craft picked up speed again, now pulled in by the Moon. Once in lunar orbit, Armstrong and Aldrin separated the lunar module, nicknamed *Eagle*, fired its descent stage, and began the landing maneuver. Collins remained in the command module, orbiting the Moon as his two fellow astronauts moved toward a date with history.

While the lander's computer controlled all maneuvers, the pilot could override the computer, and, for the final touchdown, Armstrong looked out the window and selected a level landing site. When the lander was about 5 feet (1.5 meters) above the surface, the engine shut off, and the lander touched on the Moon's Sea of Tranquility on July 20, 1969. Moments later, Armstrong radioed back his famous announcement: "Houston, Tranquility Base here. The *Eagle* has landed."

As they left the lander, a television camera mounted on the side of the lunar module sent blurred images of the astronauts back to Earth. Armstrong stepped off the pad onto the Moon and said, "That's one small step for a man, one giant leap for mankind." (According to Armstrong, most of the millions watching on television did not hear the astronaut say the word "a" before "man" because of a gap in the transmission.) Later on, he told Aldrin, who had joined him on the surface of the Moon, "Isn't it fun?"

It was 2,974 days after President Kennedy's pledge. The two astronauts planted an American flag and left behind a plaque that read:

HERE MEN FROM THE PLANET EARTH
FIRST SET FOOT UPON THE MOON
JULY 1969 A.D.
WE CAME IN PEACE FOR ALL MANKIND

The second flight to the Moon was as successful as the first, and *Apollo 12* made a precision landing on the lunar surface on November 19, 1969. Astronauts Charles (Pete) Conrad Jr. and Alan L. Bean walked to a landed space probe, *Surveyor 3*, and retrieved samples for study.

VOICES OF THE UNIVERSE:
Apollo astronaut JOHN L. SWIGERT JR.

OK, Houston, we've had a problem.

Was the film *Apollo 13* accurate?

The flight of *Apollo 13* was supposed to result in the third lunar landing, but nearly ended in the disaster memorialized in Ron Howard's historically accurate film, *Apollo 13*. The April 1970 flight became a desperate rescue mission to save the lives of three astronauts—James A. Lovell Jr., Fred W. Haise Jr., and John L. Swigert Jr. During the approach to the moon, one of two oxygen tanks that provided both breathing oxygen and fuel for the electrical power systems of the command and service modules exploded, disabling the remaining tank. Moments later, Swigert reported, "OK, Houston, we've had a problem." (In the movie, it was delivered as "Houston, we have a problem.")

Realizing that the astronauts probably did not have enough oxygen and battery power to get them back to Earth, flight controllers at Mission Control in Houston quickly ordered the crew to power up the lunar module, which had its own power and oxygen supplies but was not designed to support three astronauts. The crew then shut down the command and service modules, saving their power supply until power would be needed for descent to Earth. Using only minimal electric power during the three-day return trip to Earth, all three of them survived.

After investigating the *Apollo 13* accident, NASA determined that wires leading to a thermostat inside the tank had been improperly tested. During the flight, the wires short-circuited, causing a fire in the pure oxygen environment of the tank, resulting in the explosion. NASA redesigned the command and service modules.

Apollo astronauts landed on the Moon six times between 1969 and 1972. The last lunar mission, *Apollo 17*, landed in the Taurus Mountains on December 11, 1972. Eugene A. Cernan and Harrison H. Schmitt, whose doctorate in geology made him the only scientist to go on an Apollo mission, rode to the surface on this mission. The two of them set the lunar record for distance driving, tooling around twenty-one miles of the lunar surface on one of the golf cart–like lunar roving vehicles.

In the Soviet Union, the "party line" was that there had never been a Soviet equivalent to the Apollo program. But in the late 1980s, the Soviet Union began to release information indicating that the Soviet government actually had an ambitious lunar program that failed. Game, set, and match in the race for the Moon.

The Apollo expeditions achieved President Kennedy's stated goal of demonstrating U.S. technological superiority; the race to the Moon ended with a hands-down U.S. triumph. Was it worth it? NASA put the final cost for Apollo at $25 billion, with another $4.5 billion for the combined costs of the earlier Mercury, Gemini, and other manned programs. Looking back, it sounds like a bargain. But the space program has always had its critics. There are those who think any space program is a waste of taxpayer dollars, as well as those who think that space research should be carried out in an entirely different manner, with an emphasis on unmanned exploration that reduces the risk to human lives. To that, Apollo's defenders say the missions provided data which would have been impossible to gather through the use of probes alone. In addition, there is the "space dividend" (see below). Apollo helped industry develop new technologies that were later applied to more ordinary tasks. Significant advances in microelectronics and medical monitoring equipment that save lives, and have improved health for millions of people back on Earth, were clearly derived from the successes of the Apollo program.

But, finally, it may come back to the human spirit. Was Lewis and Clark's expedition worth it? Was Magellan's circumnavigation worth the life of the navigator? These are the extraordinary achievements that require daring and courage that add to the human experience. It is important to remember the time in which Apollo happened. The sixties were the best of times and the worst of times, in so many ways. The president who issued the challenge to go to the Moon was dead by an assassin's hand. In 1968, the Reverend Martin Luther King and President Kennedy's younger brother, Senator Robert Kennedy, were also assassinated. The country was going through the nightmare of Vietnam and the social unrest that is usually and inadequately summed up as "the sixties." Racial strife was tearing America apart. Apollo was, if nothing else, a restorative moment for the nation's nerves. Even if there were critics of the program, they were muted in the face of the extraordinary sight of the Earth as seen from the Moon. The practical impact of the Apollo missions—the space program itself—can be weighed and debated. But the human spirit, if history is any guide, demands that humanity continues to push out, breaking new boundaries. And doing it, ideally, in a way that builds on the lessons of the past.

In the end, it is difficult to argue with Space Age historian William E. Burrows's assessment; he called Apollo the "greatest human adventure; the *Odyssey* of the millennium."

Who were the "Mercury 13"?

You've heard of John Glenn, Neil Armstrong, Alan Shepard, and all the other famous astronauts. But have you every heard of Jerrie Cobb?

One of the highest-scoring members of the early astronaut training program, Jerrie Cobb was "washed out" by NASA. Jerrie Cobb didn't have the "right stuff." Literally. Jerrie Cobb and twelve other astronaut trainees lacked some of the basic equipment that the other first American astronauts had. They were women.

One of the forgotten chapters in the history of the space race is the existence of a group of women astronauts-in-training. Selected in 1959, America's first female space fliers had undergone the same screening process as the *Mercury* 7 men. They were all in excellent health, held college degrees, and were experienced pilots. Among the thirteen was Jerrie Cobb, who had been flying since she was twelve years old. One of the world's top pilots, with four world records for speed, distance, and altitude, Cobb was chosen to be the first woman to take the astronaut-qualifying tests. She not only passed the tests, but actually ranked in the top 2 percent of the astronauts in the training program. Political pull apparently did not matter either for the "ladies" of the *Mercury* 13. One member of the group was Jane Hart, the wife of Michigan senator Philip A. Hart. In addition to her airplane experience, she had been the first person licensed to fly helicopters in her state. A mother of eight, Hart also served to underscore the point that anyone who can handle labor has the "right stuff" for the stresses of space, and that if men gave birth there would probably be a national monument to Delivery Room Heroes. In face, there were some people who believed that the women, smaller and lighter than their male counterparts, would have been better candidates for space, given the lift capabilities of the early generation of rockets.

But in 1960, NASA killed the women's chances. A last-minute rule change was made, guaranteeing that women would not challenge their male counterparts. Officially, they were ruled out because they had no experience as test pilots flying jets, which were essentially military aircraft in which women could not fly at that time. Sexism in NASA typical of the late 1950s America, fears of a publicity disaster should a woman die in an accident, and the objections of the male astronauts who didn't want women competing with them for glory were really behind the decision. The Mercury 13—K. Cagle, Jerrie Cobb, Jan Dietrich, Marion Dietrich, Wally Funk, Jane Hart, Jean

Hixson, Gene Nora Jessen, Irene Leverton, Sara Ratley, B. Steadman, Jeri Truhill, and Rhea Woltman—was disbanded in 1960. Three years later, Valentina Tereshkova, a twenty-six-year-old Russian textile worker, was sent into space in another public relations coup for the Soviet space program. It would take another twenty years before Sally K. Ride would become America's first woman in space when she rode the space shuttle in 1983.

What happened to the space shuttle *Challenger*?

If the Apollo program represented the pinnacle of human achievement and the greatest moment for NASA, its lowest had to come in an awful moment on the cold morning of January 28, 1986. The tenth launch of the space shuttle *Challenger* was scheduled as the twenty-fifth space-shuttle mission. Francis R. (Dick) Scobee was the mission commander. The five other regular crew members were Gregory B. Jarvis, Ronald E. McNair, Ellison S. Onizuka, Judith A. Resnik, and Michael J. Smith. But this mission was different; the crew included Christa McAuliffe, a high school teacher from Concord, New Hampshire, mother of two and winner of a contest to become the first "citizen passenger" in space. The choice of McAuliffe was part of NASA's usually unerring sense of perfect pitch for public relations. Their understanding of the weather, engineering, and physics should have been so flawless.

The space-shuttle program, which had once again captured flagging American enthusiasm for space, had become rather humdrum in public opinion. Sending shuttles up had become as predictable as airlines' flights. But, like airlines, NASA could also fall behind schedule. By 1986, it was way behind schedule, and way over budget. Congressional budget hawks were looking for targets, and NASA and the space program had become a sitting duck, a bloated bureaucracy, since the heady days of the sixties when they had their way. As had become annoyingly routine with shuttle flights, this *Challenger* mission had been set back by several launch delays. Despite warnings from representatives of the shuttle's builders, NASA officials overruled the concerns of engineers and ordered a liftoff at 11:38 A.M. with the Florida temperature at 36°F. Seventy-three seconds into the flight, with the nation watching its first "Mom and Apple Pie" shuttle astronaut, *Challenger* disintegrated into a ball of fire at an altitude of 46,000 feet (14,020 meters). Among the stunned audience watching the launch had been McAuliffe's parents and sister. Television cameras grippingly caught their faces going from excitement to bewilderment to disbelief in a matter of seconds as it became apparent that something had gone terribly wrong.

The *Challenger* disaster was the blackest day in a history that went back to the 1950s and the 1960s, when aviation designers first began to develop the concept of winged rocket planes that could land on ordinary airfields. The idea was simple on the face of it: to make a cheaper, reusable spacecraft that could blast off like a rocket but land like an airplane. Designed to be launched over a hundred times, they were going to be America's "pickup trucks" in space, reliable work vehicles that would take America's space efforts to the next level of exploring and developing space for practical human use. NASA began to develop a reusable space shuttle while the Apollo program was still underway, and, in 1972, President Richard M. Nixon signed an executive order that officially started the space-shuttle project.

As designed, the space-shuttle system consisted of three components: an orbiter, an external tank, and two solid rocket boosters. The nose of the winged orbiter housed the pressurized crew cabin, which resembled an airplane's cockpit. Pilots could look through the front and side windows. The external tank attached to the orbiter's belly and contained the liquid propellants used by the main engines. Two rocket boosters, containing solid propellants, were attached to the sides of the external tank. After launch, these boosters, two long cylindrical tubes, dropped away, fell into the ocean, and were recovered to be reused.

In 1977, NASA conducted flight tests of the first space shuttle, *Enterprise,* named after the fictional starship from the original *Star Trek* series following a huge write-in campaign by "Trekkies," the rabid fans of the series. While NASA wasn't particularly happy about naming one of its craft after an icon of pop culture, it still worked well with the media and the public. The orbiter was piggy-backed onto a modified 747 jumbo jet for its testing phase.

The shuttle's first orbital mission began on April 12, 1981. That day, the shuttle *Columbia* was launched, with astronauts John W. Young and Robert L. Crippen at the controls. The fifty-four-hour mission went perfectly. Seven months later, the vehicle made a second orbital flight, proving that a spacecraft could be reused. Although the first four shuttle flights each carried only two pilots, the crew size was soon expanded to four, and later to seven or eight. Besides the two pilots, shuttle crews grew to include "mission specialists," experts in the scientific research to be performed, and "payload specialists," a term that was supposed to mean experts in the operation of the shuttle but which grew increasingly ambiguous to include a variety of passengers like a U.S. senator and a Saudi prince whose presence on board was more ceremonial than scientific. These were the "citizen passengers," a group NASA eventually expected to include journalists and artists. On that *Challenger* launch was the first such "citizen passenger," schoolteacher Christa McAuliffe, who would deliver classroom lessons from space in a carefully calculated measure of NASA's public relations blitz.

In the wake of the disaster, all shuttle missions were halted while a special commission appointed by President Reagan determined the cause of the accident. Its fourteen members were led by former Secretary of State William P. Rogers, who made it clear from the outset that NASA would emerge unscathed, at least publicly, and included Apollo hero Neil Armstrong and America's first woman in space, Sally Ride. In June 1986, the commission reported that the accident was caused by a failure of an O-ring in the shuttle's right solid rocket booster.

An O-ring? Think of the rubber ring that seals the top of a Mason jar of preserves. More sophisticated versions of these rubber rings sealed the joint between the two lower segments of the booster. The dramatic high point of the hearings came when committee member Richard P. Feynman, a veteran of the Manhattan Project and one of the country's most prominent physicists, dipped a piece of O-ring into a glass of ice water and used a vice clamp acquired from a hardware store to show that the cold rubber was brittle. Design flaws in the joint and unusually cold weather during launch caused these O-rings to fail, a possibility that was apparent in test reports from one of the shuttle's contractors.

That was the physical cause of the disaster but the ultimate cause was the decision to hurry a launch to justify the shuttle program which was falling farther and farther behind schedule and costing more than it was ever predicted to cost. The fact that NASA wanted to keep Christa McAuliffe's classroom lesson plan intact also played a role in the decision to launch, as did President Reagan's schedule. He was supposed to deliver the State of the Union Address that same night and planned to refer to the shuttle and McAuliffe. Many critics believe that there was White House pressure applied to NASA to keep to the schedule.

As teachers like Christa McAuliffe always tell their students in elementary school, "Haste makes waste."

VOICES OF THE UNIVERSE:
WILLIAM E. BURROWS, *This New Ocean*, 1998

Challenger was lost because NASA came to believe its own propaganda. The agency's deeply impacted cultural hubris had it that technology—engineering—would always triumph over random disaster if certain rules were followed. The engineers-turned-technocrats could not bring themselves to accept the psychology of machines without abandoning the core principle of their own faith: equations, geometry, and repetition—physical law, precision design, and testing—must defy chaos. No matter that astronauts and cosmonauts had perished in precisely designed and carefully tested machines. Solid engineering could always provide a safety margin, because the engineers believed, there was complete safety in numbers. That made them arrogant. (p. 560)

Cell phones and Speedos:
What have you done for me lately, NASA?

From hospital beds to ski slopes and swimming pools, Space Age technology has improved life on Earth, the fringe benefit of the space program. While critics over the decades have derided spending on space exploration when there are so many unfinished jobs on Earth, the space program has actually been an investment with tremendous dividends for people everywhere. Even in the heyday of Apollo, the cost of the space program was said to work out to about 50 cents per American. That is, at least, what the NASA publicity machine says. NASA sees itself as a very large stone dropped in a pool. The dollars spent on research, equipment, and salaries devoted to the space program send out ripples that benefit the entire economy. By one NASA estimate, for every tax dollar spent on the space program, seven dollars' worth of new business is generated in the U.S. economy. A shuttle mission these days costs in the neighborhood of $500 million. Obviously, satellites do more than look at space—they deliver improved weather forecasts that can save lives and salvage harvests, they keep your cell phone working, and bring you the Olympics live!

EARTHLY DIVIDENDS OF THE SPACE SYSTEM:

The following list highlights some of the many practical and life-saving inventions and technology that have been adapted from NASA technology developed for use in the space program.

- Smoke detectors invented for use on Skylab are now used in millions of homes worldwide.

- Laser eye-surgery comes from satellite atmospheric studies.

- Radiation blockers for sunglasses screen out harmful waves to enhance eye protection and reduce eye damage.

- Pump therapy to control injections of drugs for diabetes derived from satellite technology.

- Tempur, a material for mattresses that relieve back pain, derived from technology developed to reduce gravity forces experienced during takeoff.

- Thermal insulation used in clothing, fire protection equipment, survival gear, toxic chemical protection.

- Neuromuscular electrical stimulation to assist spinal-injury victims.

- Ocular screening system makes eye tests easier—early detection helps prevent blindness in children.

- Improved prosthetic devices from a design that was meant to help keep the shuttle's fuel tanks supercold.

- Magnetic Resonance Imaging (MRI)—developed from remote sensing devices on satellites.

- Exercise machines developed for astronauts during extended periods in space when weightlessness and other conditions have been found to affect many aspects of the body.

- Light therapy aids sufferers of Seasonal Affective Disorder—a condition caused by lack of sunlight—developed to help astronauts with the sleep-wake cycle.

- Fire scanners that detect sources of visible and invisible flame as well as find unconscious people in smoke-filled rooms.

- Search-and-rescue technology has saved more lives than any other application from the Space Program—developed from satellite programs for atmosphere and climate monitoring.

- Satellite imagery has been used to detect major archeological finds such as the location of ancient civilizations in Costa Rica and the five-thousand-year-old Mesopotamian city of Ubar in modern-day Iraq.

- Satellite map resources that can provide vital data about food and water conditions.

- Save the dolphins! A device developed for spacecraft location is now attached to large fishing nets called "gill nets." The device emits a warning signal audible to dolphins, who can then avoid being caught in the nets.

- Computer-enhanced imaging that will improve brain scans and allow 3-D body imaging.

- Computerized solar water-heating systems, based on temperature control systems used on Skylab.

- Accurate geological surveys detect resources or pinpoint leaks in pipelines.

- Infrared mapping for urban planning.

- Plants that purify sewage—protein-rich water hyacinths are cultivated for use as purifiers in sewage lagoons.

- Electrolytic water filters designed for spacecraft are the foundation of domestic water purifiers.

- Weather-resistant coatings—one was used to protect the surface of the Statue of Liberty.

- Running shoes adapted from spacesuit technology.

- The search for ways to reduce aircraft fuel consumption resulted in silicon ribbing for racing swimsuits. Speedos!

- Anti-fogging spray, designed after an astronaut was put out of commission, is finding a wide variety of uses on windows, camera lenses, and helmets for pilots and motorcyclists.

- The space race and the computer age coincided and have developed side by side. NASA is involved in the mission to create the world's biggest supercomputer, which will link industry, government, and academic users around the country.

- Ergonomic seats designed for astronauts are being developed into chairs for the elderly, support chairs for paraplegics, and desk chairs that reduce work-stress injuries.

- Robotic manipulators with industrial applications may aid disabled and paraplegic people in everyday tasks.

- Computer-chip development in a totally dirt-free environment.

- Improved design efficiencies, which will create products that conserve energy and pollute less.

- Water-jet coating strippers designed to clean the shuttle's reusable rocket boosters are replacing toxic chemical strippers.

- Highway traffic control systems that will help manage urban congestion and improve road safety.

This list was adapted from David Baker's *Inventions from Outer Space*. New York: Random House, 2000.

The next great phase in space development and its practical consequences will come as the private sector expands its role into what once was NASA's exclusive turf, as business learns that space exploration can pay actual dividends. For instance, late in 1999, two firms, Lockheed Martin Astronautics and SpaceDev, announced that they would work to develop and market lower-cost space access for small payloads. Just as Federal Express came along and transformed the U.S. Postal Service, private space developers are expected to compete with NASA for the business of putting payloads for profit into space.

In the November/December 1999 issue of *Ad Astra*, the magazine of the National Space Society, an organization devoted to space exploration, executives of SpaceDev held open the possibility of taking orders for Mars, offering to deliver small payloads on Mars entry trajectories for a fixed price of about $24 million. The estimated NASA procurement cost for a similar mission is thought to be significantly higher than the SpaceDev fixed price. During a convention of the Mars Society, SpaceDev Chairman Jim Benson said, "Inner planet missions like this one can now be done for about the cost of a private jet."

Many of those plans also involved the burgeoning telecommunications industry which shot ahead—along with the Nasdaq stock markets, in the decade of the 1990s. With a "cell phone in every pot and two satellite dishes in every yard," space was increasingly being viewed not as the metaphorical "new frontier," but the next jackpot. But as the third millennium opened, some of those grandiose plans were coming in for a cold shower. A venture called Iridium, which had planned to hoist dozens of satellites into space, one of the key players in this potential multibillion-dollar industry, went bust. Iridium filed for Chapter 11 after failing to sign up enough subscribers for a telephone satellite system that cost more than $5 billion. The bankrupt Iridium was planning to let the satellites they had launched fall back to Earth, but the Pentagon agreed to spend $72 million for two years of satellite service from Iridium and Boeing agreed to operate the satellite system under contract. For the moment, the future of space as the playground of venture capital had come back down to Earth.

Most space visionaries believe that the future of space exploration can only be driven more by commerce than scientific curiosity or military needs. Edwin "Buzz" Aldrin, the second man to set foot on the Moon, is one of them. So is Arthur C. Clarke, the science-fiction visionary behind *2001: A Space Odyssey*. In 2001, appropriately, Aldrin founded ShareSpace, a nonprofit organization whose mission is to open space travel to the public. At the time, he predicted that space-adventure travel would become a reality in ten to twelve years. Clarke, an enthusiastic proponent of space colonization for decades, agreed with Aldrin but saw a more practical aspect in the potential for space. In Clarke's vision, the colonization of other planets is the only way for the human race to survive the crowding and degradation of the Earth's environment, or as a possible alternative to Earth if a catastrophic collision ever became likely.

During the Cold War period, space was the exclusive province of government—essentially, just two governments at that. The end of the Cold War, the demise of the Soviet Union, and post-Communist Russia—which recently auctioned off much of its antiquated space equipment in a desperate tag sale for cash—have opened the way for commercial ventures to take the lead. And other countries are stepping into the gap. Today, privately funded commercial spending on space ventures is greater than government spending. The Cold War is over and the privatization of space has begun. There are Golden Arches in Moscow and Beijing. It may not be long before they are casting their glow on the Sea of Tranquility.

Back to the future: When John Glenn returned to space, was it good science or just a publicity stunt?

The space shuttle resumed flying on September 29, 1988, with the launch of the redesigned shuttle *Discovery*. During the next few years, many long delayed missions were carried out. The *Galileo, Magellan,* and *Ulysses* space probes were launched. Large scientific research satellites such as the Hubble Space Telescope, the Compton Gamma Ray Observatory, the Upper Atmosphere Research Satellite, and the Chandra X-Ray Observatory were placed into orbit. Spacelab, a working science laboratory that operated inside the payload bay of the shuttle, was utilized to study astronomy and space medicine.

And spacecraft from the United States and Russia resumed joint operations in 1995, twenty years after the Apollo-Soyuz mission. On June 29, after three years of negotiations, planning, and practice missions, the space shuttle *Atlantis* docked with Russia's Mir Space Station. *Atlantis* carried a replacement crew of Russian cosmonauts to Mir and brought the station's former crew home to the earth. Among the returning crew members was astronaut Norman Thagard, who had ridden a Russian rocket to Mir on March 14, 1995. Thagard had spent 115 days in space, breaking the previous U.S. record of 84 days. On March 22, 1995, three cosmonauts who were on Mir when Thagard arrived made their return voyage to Earth. They included Valery Polyakov, who set an international record of 438 days in space. On July 15, 1996, astronaut Shannon Lucid, aboard Mir, broke Thagard's record by spending her 116th consecutive day in space, and on September 7, 1996, Lucid broke the women's record of 168 consecutive days in space, a record that had been set by cosmonaut Yelena Kondakova in 1995. On September 26, Lucid returned to Earth aboard *Atlantis,* having spent 188 consecutive days in space.

But few of these achievements attracted the level of international attention that came back to the shuttle when *Discovery* was launched in October 1998. Aboard was U.S. Senator John Glenn, America's Golden Boy of the Space Age, first American to orbit Earth. At age seventy-seven, he became the oldest person to fly in space, and the question of space and the aging process were NASA's publicly stated reasons for his return. The "scientific" aspects of Glenn's trip took a hit before he left the ground, when he was pulled out of one of the experiments in which he was supposed to participate. Even supporters of the Glenn return jaunt were cynical about NASA's justification. As *The New York Times* noted in an editorial, "The stated justification for this excursion is mostly nonsense, a fig-leaf rationalization dreamed up by Mr. Glenn during two years of lobbying for the trip. . . . There is hardly a burning need to study the effect of weightlessness on an elderly man's physical process. . . . Some critics suspect that this new ride is a payoff to Mr. Glenn for his yeoman work in parrying attacks on President Clinton. . . . But the simpler explanation is that NASA administrators see immense value in The Return of John Glenn" (*New York Times*, January 17, 1998).

The *Times* had that one right. If "Geriatrics in Space" was what they were really looking for, one thing was certain: NASA had not forgotten how to get the front-page headlines. The Glenn mission, which went off without difficulty, was the most widely covered space mission in many years. While much of the coverage had a snicker and a derisive titter about it, and late-night television had a plum of a target for

©Steve Sack/Star Tribune

several days, Glenn's triumphant return had the agency singing "Happy Days Are Here Again." NASA's chief administrator Daniel Goldin said at the time of Glenn's return that decisions about future shuttle missions including elderly crews would come in a year or two. By early 2001, no such announcements had been made. But memories can be short. Even though people may not have completely forgotten the *Challenger* disaster of twelve years earlier, NASA seemed to have rediscovered the magic that came from doing things right once again.

If you break a mirror in space, is it seven years of bad luck?

With the tragic memory of *Challenger* fading as the revamped shuttle program forged ahead, NASA was still under heavy political pressure. More than ever before during the 1980s, under the Reagan Administration, it had to justify its missions and their costs. Conceived in the 1960s, the Large Orbital Telescope, as it was first called, was designed to be the largest, most complex and powerful observatory ever sent to space. One wit suggested Congress might find it more appealing if they called it the Great Optical Device, because then Congress would be funding GOD. Instead, the space telescope was named the Hubble Space Telescope (HST) in 1983, in honor of the American astronomer who had proved that the universe is expanding.

Launched in 1990, this extraordinary piece of equipment quickly became another huge NASA embarrassment, although not a tragic one as with the *Challenger* disaster. Choosing a low bidder to build what turned out to be a primary mirror, NASA discovered one month after launch that there was a flaw in the shape of its main mirror. The words "You get what you pay for" were more than whispered around NASA water coolers. The mocking and criticism were heard around the world. As television's Jay Leno chimed in, "Hubble is working perfectly. But the universe is all blurry."

The mirror had to be corrected. In 1993, as part of a planned servicing-and-instrument-upgrade mission, NASA astronauts aboard the space shuttle *Endeavour* gave the Hubble a new pair of glasses. During a spacewalk, astronauts removed a faulty camera and added some fairly costly new lenses.

But if you ask any astronomer or space scientist, it was worth it. Hubble began to surprise even its harshest critics. Looking to the very edges of the known universe, Hubble showed people on Earth galaxies in formation, some *40 billion* galaxies. In December 1995, the Hubble Space Telescope was trained on an "empty" area of sky near the Big Dipper, now termed the Hubble Deep Field. Over time, some 1,500 galaxies, mostly new discoveries, were photographed. In May 1997, three months after astronauts installed further new equipment, U.S. scientists reported that Hubble had made extraordinary findings—including evidence of a black hole 300 million times the mass of the Sun, located in the middle of galaxy M87 in the Virgo cluster.

While Hubble has revolutionized astronomy in expanding human vision to distant galaxies, it is still working within constraints. In 2009, Hubble's successor is supposed to take the next step. The Next Generation Space Telescope will be designed to carry a reflective surface ten times greater than Hubble's but at a fraction of the weight. As planned, this mirror will be collapsed, like a flower petal, at launch, and open up when it reaches its destination, a point in orbit nearly one million miles away from Earth, a place at which the gravity of the Sun, Moon, and Earth balance each other, but well beyond the reach of the shuttle. In other words, there won't be any handyman specials for the NGST. It has to work right from day one.

Space, toilets, and sex: How will the space station work?

In 1982, President Ronald Reagan, perhaps in a very deliberate echo of John F. Kennedy nearly a quarter of a century earlier, publicly urged the nation to build a "more permanent presence in space." Two years later, Reagan authorized the go-ahead on construction of a large, permanent space station "within a decade" and invited other nations to join in the effort. In an era of tremendous budget deficits, the space station was NASA's intensive-care patient—always on the precipice of having its life support unplugged by a Congress looking to cut dollars from the space program and in-fighting with the military that didn't want the thing built at all. Typically, the program was often saved by powerful legislators whose districts included key contractors for the space station's components. (Just in case you have ever wondered why the Space Command is in Texas, Lyndon B. Johnson was a very powerful senator before he became JFK's vice-president. A Houston command center was a legacy to LBJ's budget prowess.)

As designs for the new station changed and the estimated costs skyrocketed, congressional patience with an agency that had already damaged its credibility after *Challenger* was severely tested. The promised completion date kept slipping. In 1993, President Bill Clinton directed NASA to rethink the proposed space station to reduce both the cost and construction time, and the United States, Canada, Japan, Russia, and the European Space Agency became partners in a program to build a redesigned space station. The International Space Station would be built from several pressurized modules and solar-power panels, delivered separately by shuttles and constructed in space. Even that decision was controversial, because it was possible to construct the space station and launch it as a single unit.

At the dawn of its actual operation in early 2001, the International Space Station was slated to become the largest multinational scientific project in history, with sixteen nations cooperating: United States, Russia, Canada, Belgium, Denmark, France, Germany, Italy, Netherlands, Norway, Spain, Sweden, Switzerland, United Kingdom, Japan, and Brazil. When completed, the station will have six laboratories with living space for up to seven people and be powered by an acre of solar panels.

Like some genius child's Lego creation in space, the station got under way in November 1998, as the Russian-built *Zarya* control module was launched by rocket—the first step in the assembly of the station. On December 4, 1998, the U.S.-built *Unity* connecting module was launched on the shuttle *Endeavour*. The shuttle crew attached *Unity* and *Zarya*. In May 1999, *Discovery* brought supplies in the first docking with the International Space Station, now named "Alpha"—the first. In November 2000, two Russian cosmonauts, Yuri Gidzenko and Sergei Krikalev, and American astronaut Bill Shepherd, became Alpha's first full-time crew. In February 2001, the American-built Destiny lab was delivered by shuttle and successfully linked to the space station, expanding its working areas and its laboratory and computing capabilities. The demands of the space station will call for the greatest logistical challenge in history—regularly supplying and restocking the station. That will be done through the regular arrival and departure of three Italian-built modules—no doubt inspiring many jokes about whether the Italians will deliver pizza. These modules are scheduled to arrive and dock at the station, bringing fresh food, equipment, and science experiments, and then return to Earth carrying waste and completed experiments.

The space station will offer the best in creature comforts that space has yet to offer. And the most frequently asked questions at NASA still involve the toilet and other bodily functions. Even President Harry Truman, at the dawn of the Space Age, was curious. When shown a *Mercury* capsule, he reportedly asked, "How do these guys take a leak?" That technology has evolved since the dark ages of *Mercury* when the first-generation astronauts essentially wore heavy-duty diapers. Maximum Absorbency Garments (MAG) are still worn during spacewalks and shuttle launches. Presumably, astronauts, like small children before a car ride, are reminded not to drink too much and make sure they "go before we leave." The new generation of space toilets operate like giant vacuum cleaners, since they don't use water to flush, sucking solid waste into a commode. They come with loops and restraint bars to prevent the user from floating off. Liquid waste is collected through a waste collector, and the urine is stored to be tested upon return to Earth.

As for sex in space, it also is a question frequently asked of NASA, and as extended coed stays in space become more routine, is a legitimately logical one. So far, nobody's talking and NASA has nothing to say on the subject.

MILESTONES IN THE UNIVERSE
1970–2001

1970 *Apollo 13*: A third lunar mission, with a crew of James A. Lovell Jr., Fred W. Haise, and John L. Swigert, nearly turns disastrous; the lunar landing is aborted because of equipment failure. The service module explodes fifty-five hours into the mission; the crew limps home using the lunar module as a lifeboat.

Japan and China launch their first satellites.

The first "jumbo jet," the Boeing 747, goes into transatlantic service.

1971   Alan B. Shepard Jr. and Edgar Mitchell collect 98 pounds of moon rocks during the *Apollo 14* lunar mission; pilot Stuart A. Roosa orbits above the Moon.

Three Soviet cosmonauts dock their *Soyuz 10* craft with Salyut 1, the first space station.

During the *Apollo 15* mission, David R. Scott and James B. Irwin drive a golf cart–like lunar rover on the Moon's surface.

In the first orbit of another planet, the American *Mariner 9* enters orbit around Mars; the craft eventually returns 7,329 pictures of Mars.

The Soviet space probe *Mars 3* lands on Mars and sends back an unreadable television signal before it goes dead.

The British launch their first satellite.

Intel introduces the first microprocessor, now known as the "chip."

1972   The *Apollo 16* crew lands on the Moon in April and takes three moonwalks. (Ken Mattingly, in the lunar orbiter, makes the longest solo U.S. flight.)

Pioneer 10, a U.S. space probe, is launched; in 1983, it becomes the first man-made object to leave the solar system.

The last manned lunar landing is launched on December 7; two astronauts spend forty-four hours on the surface of the Moon.

1973   The first of three Skylab missions launched in this year. Crews will conduct medical and scientific experiments and obtain data for extended space flights.

1974   A Soviet space probe lands on Mars.

Mariner 10 makes the first fly-by of Mercury.

1975 The first pictures from the surface of Venus are received from Soviet *Venera* space probes.

In the first U.S.-Soviet space cooperation, the Apollo-Soyuz space project is launched; a three-man Apollo crew docks with a two-man Soviet crew.

The first liquid crystal display (LCD) for calculators and digital clocks is marketed.

The first personal computer, the Altair 8800, is introduced.

1976 The *U.S. Viking* space probes soft-lands on Mars and begins sending back pictures; it functions for three and a half years.

The Concorde becomes the first supersonic airliner to operate regular passenger service.

1976 U.S. space probes *Voyager 1* and *Voyager 2* are launched on missions to Jupiter and the outer planets; *Voyager 1* passed near Saturn in 1980.

Apple introduces the first personal computer available in assembled form; the following year, Apple releases the first disk drive for use with personal computers.

1979 Skylab, an early American space station that had been abandoned, falls into Earth's atmosphere as a result of an intense solar wind caused by increased sunspot activity. After intense media coverage of Skylab's possible crash landing, it breaks into pieces that land harmlessly in western Australia.

First exploration of Saturn by *Pioneer 11*; its transmissions ended in 1995 in the outer solar system.

1980 The VLA—Very Large Array—radio telescope begins operation in Socorro, New Mexico.

Voyager 1 and *Voyager 2* both fly by Saturn providing much information about the planet, its moons, and its ring system.

1981   University of Wisconsin researchers discover the most massive star known; R136a; it is 100 times brighter than the Sun and has 2,500 times the Sun's mass.

Maiden flight of the space shuttle. On April 12, the first flight of STS-1, the Space Transportation System's *Columbia*. It proves that the concept of a reusable space shuttle is viable. *Columbia* makes a second mission—the first reuse of a spacecraft.

IBM introduces its personal computer (PC) using what will become the industry-standard disk operating system (DOS).

1982   *Columbia* makes two more trial flights and, on November 11, is launched for its first operational mission; it deploys two communications satellites.

Compact disk players are introduced; Compaq introduces the first "clone" of an IBM personal computer.

1983   On June 18, the second U.S. space shuttle, *Challenger*, is launched. With the first five-passenger crew, it carries the first American woman astronaut, Sally K. Ride. On its third flight, in August, *Challenger* carries the first African-American astronaut, Guion Bluford, Jr., into space.

Apple introduces the computer mouse and pull-down menus; in the next year, it introduces the Macintosh.

1984   During *Challenger*'s fifth mission, two astronauts use Manned-Mission Maneuvering Units (MMU), or jet packs, to make the first untethered space walks.

The *Soyuz 10* mission becomes the longest crewed mission to date, and two cosmonauts make a record six spacewalks.

1985   Construction of the Keck Telescope, the world's largest, begins on the island of Maui, Hawaii.

Aboard *Discovery*, Senator Jake Garn becomes first U.S. senator in space.

Aboard *Discovery*, Salman al-Saud becomes the first Arab in space.

Challenger carries the first eight-person crew.

Aboard *Atlantis*, Rodolfo Neri becomes the first Mexican in space.

Aboard *Columbia*, Bill Nelson becomes the first U.S. representative in space.

1986 On January 28, the space shuttle *Challenger* blows apart seventy-three seconds after launch. Six astronauts—Dick Scobee, Mike Smith, Judith Resnik, Ronald McNair, Ellison Onizuka, and Gregory Jarvis—and schoolteacher Christa McAuliffe are killed in the disaster.

U.S. space probe *Voyager 2* passes close to Uranus.

The Soviet Mir (Peace) Space Station is launched in February; it becomes the first permanently manned space station.

First encounter of the coma of a comet by *Giotto*.

1987 Two members of the Soyuz crew set a new space duration record of more than a year.

1988 Redesigned shuttle *Discovery* makes first flight since *Challenger* disaster.

1989 In May, the space shuttle *Atlantis* is launched. In the first deployment of a planetary spacecraft from a crewed spacecraft, the crew deploys the *Magellan* on a journey to orbit Venus.

In October, space shuttle *Atlantis* deploys a Jupiter orbiter, *Galileo*, which uses Earth's gravity to propel it towards Jupiter.

1990 In April, the shuttle *Discovery* deploys the Hubble Space Telescope (HST) and reaches a record shuttle altitude of 319 miles (about 520 kilometers); later that year, *Discovery* deploys the *Ulysses* spacecraft to investigate interstellar space and the Sun.

1991 The *Galileo* probe takes the closest-ever picture of an asteroid, Gaspra. Later, *Galileo* becomes the first craft to enter Jupiter's atmosphere and orbit Jupiter. Communication with *Galileo* is lost in 1993, after probes are deployed to Jupiter's surface.

1992   The shuttle *Endeavour* retrieves a satellite and reboosts it into orbit; three crew members make a record-breaking eight-and-a-half-hour spacewalk.

An *Endeavour* mission in September includes two space firsts: the first African-American woman, Mae Carol Jemison, and the first married couple, Jan Davis and Mark Lee.

1993   In December, *Endeavour* makes a service-and-repair mission to the Hubble Space Telescope. New spacewalk duration record set: 29 hours 40 minutes.

1994   Aboard *Discovery*, Sergei Krikalev becomes the first Russian cosmonaut on a U.S. shuttle.

1995   The *Discovery* makes a rendezvous with Mir. Eileen Collins is the first woman to pilot a shuttle.

In March, shuttle data from *Endeavour* is made available on the Internet.

In June, the launch of *Atlantis* is the one-hundredth crewed U.S. flight.

In November, *Atlantis* carries a docking module to be left at Mir.

1996   *Atlantis* delivers U.S. astronaut Shannon Lucid to Mir for an extended stay of 188 days, a U.S. and woman's duration record.

Mars Orbital Surveyor begins orbiting Mars in November. A two-year mapping survey of the entire planet; observes magnetism on Mars, discovers evidence of liquid water in geological past discovered in June 2000.

1997   A Soyuz mission in February delivers a new crew to the now-troubled Mir. The crew survives a fire onboard the space station and a collision with a supply ship.

In February, the *Discovery* makes a second mission to service the Hubble Space Telescope, increasing capabilities of the HST.

Atlantis docks with Mir and delivers a new computer. During the stay, there is another collision with a cargo ship, the worst such accident yet.

In July, *Pathfinder* (launched in December 1996) lands on Mars; the rover *Sojourner* makes measurements of Martian soil and sends thousands of images of Martian surface; ceases to operate in September.

In October, *Cassini* launched to Saturn; arrival scheduled in 2004

1998   *Endeavour* docks with Mir, delivers water and cargo.

John Glenn returns to space. Launched aboard *Discovery* in October, U.S. Senator John Glenn, at age seventy-seven, becomes the oldest person to fly in space.

In December, Mars Climate Observer launched; communication lost in September 1999.

1999   *Endeavour* crew begins construction of the International Space Station. The crew attaches the U.S.-built *Unity* connecting module with the previously deployed Russian-built *Zarya* control module.

In January, Mars Polar Lander launched; communication lost in December 1999.

In February, *Stardust* launched to rendezvous with Comet Wild-2 in 2004; its mission is to gather dust samples and return them to Earth in 2006.

In May, *Discovery* crew transfers nearly two tons of supplies to the International Space Station (ISS).

In July, Aboard *Columbia*, Eileen Collins becomes the first woman to command a space shuttle. The crew deploys the Chandra X-Ray Observatory to study the distant universe.

In December, *Discovery* upgrades the Hubble Space Telescope.

2000   *Endeavour* uses radar to make the most complete topographic map of Earth's surface ever produced.

In May, *Atlantis* services and resupplies the International Space Station, boosting its altitude.

In September, *Atlantis* prepares the International Space Station for its first permanent crew.

In October, *Discovery* installs permanent framework structure on International Space Station.

Two Russian cosmonauts and American astronaut become the first working crew members onboard the International Space Station.

2001   In January, detection of X-ray emissions offers best evidence to date of an event horizon, confirming the existence of black holes.

In February, astronauts aboard space shuttle *Atlantis* deliver and attach Destiny laboratory, the chief scientific component of the ISS.

In February, after orbiting Eros asteroid for one year, NEAR spacecraft successfully lands on Eros and continues to transmit data.

Mir space station plunges into the Pacific Ocean after fifteen years in space.

In April, Dennis Tito, sixty-year-old millionaire entrepreneur, pays $20 million to become the first space tourist.

VOICES OF THE UNIVERSE:
CARL SAGAN from *Cosmos* (1980)

Our world is now overflowing with life. How did it come about? How, in the absence of life, were carbon-based organic molecules made? How did the first living things arise? . . . And on the countless other planets that may circle other suns, is there life also? Is extraterrestrial life, if it exists, based on the same organic molecules as life on Earth? Do the beings of other worlds look much like life on Earth? Or are they stunningly different—other adaptations to other environments? What else is possible? The nature of life on Earth and the search for life elsewhere are two sides of the same question—the search for who we are.

We have walked on the Moon—in spite of what sensational pseu-dodocumentaries would have you believe. We have landed roving robots on the surface of Mars and sent one careening about the red planet like a remote-controlled race car. People now work full-time in space on an orbiting space station. Our spacecraft have left the Solar System, taking Earth's technology and message where "no man has gone before." A new generation of space-based telescopes and obser-vatories has peered farther out toward the edges of the universe and helped discover new planets. We have begun to understand the very stuff of stars. Our remarkable success in exploring the heavens—par-ticularly the extraordinary accomplishments in less than half a century since humans first left Earth—will go on.

For the average person, the prospect of continuing space explo-ration may be cause for excitement, a disinterested yawn, or even out-rage over their perceived waste of tax dollars. But there is one question above all that seems to grab everyone's attention whether at the mul-tiplex cinema or the supermarket checkout line: Is there anything or anybody else out there?

Accept for a moment that, so far, the answer is a resounding "No." No Martians. No Klingons or Romulans. No Alien for Sigourney Weaver to hunt. No Wookies or Jabba the Hutt. Not even a Yoda or an Ewok. Can you imagine a universe without anything else alive? For most of humanity's history, we have imagined other creatures existing. Out there. Somewhere. For many people, it has been an article of faith: there must be other life, other civilizations. In 1820, a German mathematician wanted to clear a huge right triangle in the Siberian forest to show any passing extraterrestrials that we understood geome-try. Another astronomer once suggested setting huge kerosene fires in the Sahara Desert as a kind of cosmic "Hello out there."

The belief that alien beings exist is not an offshoot of twentieth-century space exploration. It was first popularized in the mid-nine-teenth century by French astronomer Camille Flammarion, one of the first scientists who tried to popularize scientific discovery for the average reader. A believer in forms of spirituality like reincarnation, Flammarion wrote a speculative "nonfiction" book, *Real and Imagi-nary Worlds,* in which he described plants and divinely created beings on other worlds. He was writing around the same time as the better-known Frenchman, Jules Verne, whose *From the Earth to the Moon* (1865), a work of fiction grounded in the hard facts of astronomy, profoundly influenced the first generation of rocketeers, including Goddard, Tsiolkovsky, and Hermann Oberth. But probably the most influential imaginative work of alien existence was British writer H. G. Wells's classic *War of the Worlds*, which began to appear in serialized form in Europe and America in 1897 and was published as a book in 1898. Depicting an invasion of London by terrifying Martians who eventually succumb to terrestrial disease, Wells set the standard for countless B-movies and works of pulp fiction.

In the century since Wells scared the collective pants off earthlings, belief in the idea that the vast universe is teeming with life has taken two paths. The first is the whole spectrum of "UFO-Alien Abduction-Alien Visitation-*Chariots of the Gods*" belief, a collection of pseudoscientific concepts that all share a complete and unshakable faith in the widespread existence of up-to-no-good aliens who routinely stop by Earth. Steven Spielberg's *E.T.* went a long way in showing an alternative possibility: benevolent, sophisticated aliens who find Earth "a nice place to visit but they wouldn't want to live here." Remarkably, members of both groups of aliens drop by, have a quick cup of coffee, but never decide to stay for a longer visit. These alien visitations have been the inspiration for works of science fiction ranging from *The War of the Worlds* to *The X-Files*. There is a large and highly committed group—some scientists probably think they should *be* committed—of people who believe completely in the reality of alien visitation.

There is no credible scientific proof for any such claims. Most of the so-called evidence is anecdotal ("I saw strange lights last night"), based on flawed perceptions of natural phenomena (auroras, meteors) and human invention (weather balloons, jets), or outright con jobs, such as the infamous *Alien Autopsy* documentary shown on television, which later proved to be a complete fabrication. (Needless to say, the ratings were sensational.)

Everything from the first sightings of flying saucers to the mysteries at Roswell, New Mexico (now reliably known to have been a weather balloon), to the likelihood that the infamous U.S. Air Force "Area 51" is a testing ground for the future generation of highly classified military aircraft, can be explained. But that does not deter the popularity of modern myths of UFO hunters. The fact is that many people believe in these things simply because they choose to. For most of us, aliens are intrinsically a lot more appealing, interesting, and comprehensible than quantum physics. People believe what they wish, whether it is the entertaining and provocative, such as *The X-Files* or Sagan's *Contact*, or the tabloid headlines ("Aliens Ate My Mom") that captivate us so teasingly at the supermarket checkout counter.

VOICES OF THE UNIVERSE:
"So where are they?" Physicist ENRICO FERMI
asked his colleagues on
the Manhattan Project, regarding aliens.

"They are among us," replied Leo Szilard, Hungarian physicist and one of the fathers of atomic energy. "But they call themselves Hungarians."

Has anyone been abducted by aliens?

One of the most famous bad jokes in stand-up history is Henny Youngman's "Take my wife. Please." Ba-da-boom. The revised modern version has a man saying that to a hovering spaceship. Just as America went through a period of intense UFO fascination during the 1960s, the extraterrestrial fad of the moment has shifted to alien abductions. The vast majority of investigated cases are usually unsupported by evidence (such as reliable witnesses, physical or medical evidence) and fall into a few areas of likely explanation. Many experts in this area believe that more people have begun to report these encounters because they are influenced by what they hear and read. Especially popular are *The X-Files* and Whitley Streiber's book *Communion*, a best-selling account of the archetypal abduction experience: the author describes being taken aboard a mother ship, experimented on, and subjected to strange medical tests. Women who make such reports routinely describe their impregnation by aliens, presum-

"Very nice, but I understood there were to be bizarre sexual experiments."

ably as part of a plan to colonize Earth. Streiber's book helped raise a feverish pitch in a society that seems to prefer to accept the conspiracy explanation of events over the scientific. Many of those who make these reports are working with hypnotists who may have influenced or even shaped these abduction accounts. Many "abductees" may have experienced a form of dream state that is extremely powerful—or hallucinated these experiences. Finally, there are some that are outright hoaxes or the result of mental illness.

Who is looking for life?

When it comes to extraterrestrial life, science is divided into, more or less, two camps at extremes of the spectrum—believers and cynics—with a large middle group saying, "Show me." The dismissal of most popular alien claims by mainstream scientists does not mean they do not accept the possibility of extraterrestrial life. And, as a rule, "absence of evidence does not mean evidence of absence."

On the contrary, there are significant attempts being made to answer the question of other life "out there." The scientific search for life in the universe currently takes two separate approaches. The first is to use probes and other unmanned methods of gathering scientific evidence from Mars and other locations in the solar system, those places in which life might exist, to test for the prospects of an environment that could support life both as we know it or in unexpected forms. By "life," these scientists are talking about anything from bacteria on up—not Marvin the Martian. This quest includes attempts to find stars outside our solar system with planets that are capable of supporting life, a key focus of future unmanned probes.

In another large camp are the dedicated professional scientists who are seeking out "new life" through the means of an ambitious Earth-based search of the heavens. These are the Extraterrestrial-Life Seekers, who included the late Carl Sagan and astronomer Frank Drake. These scientists have based their "faith" in extraterrestrial life on a probability factor that has been expressed in a famous mathematical concept called the "Drake Equation." Named for astronomer Frank Drake, who devised it in 1961 when he was thirty years old, it is a mathematical formula for computing the possibility of life forms in the Milky Way, which are or might be trying to contact us through radio transmissions.

$$N = R^* \times fP \times nE \times fL \times fI \times fC \times L$$

Don't be afraid! It looks more threatening than it actually is. Simply put, the Drake Equation attempts to calculate N, the number of technically advanced civilizations active in the Milky Way capable of broadcasting radio transmissions. To break it down:

R^* = the rate at which stars suitable for hosting habitable planets are formed in the galaxy each year. We do know that new stars are formed all the time. The Milky Way has 4 billion stars and is 10 billion years old, so this number is estimated at four per year.

fP = the fraction of those stars that have planets, in other words, other solar systems. Progress has been made here in identifying stars with at least fifty planets orbiting them.

nE = the number of Earth-like planets per solar system capable of incubating and supporting life. In our solar system, this number is at least one, based on Earth, with Mars a possible, but unproven, second. So far, there are none known in any other solar system.

fL = the fraction of Earth-like planets where life will develop, a highly speculative figure.

fI = the fraction of planets with life that develop intelligence, also highly speculative.

fC = the fraction of planets with intelligent life that develop the technology to communicate, again highly speculative.

L = the lifetime of a technological civilization. Highly speculative, because we have only one example to work with, our own, which has proven that it is also capable of wiping itself out.

Solving this equation can become a fascinating parlor game, because you can plug in just about any figure you like. Even Drake himself has admitted that no one knows the values of many of the factors in the equation, although some are a little more certain or predictable than others. The late astronomer and science popularizer Carl Sagan figured that the N must equal at least a million civilizations in the Milky Way alone, a highly optimistic figure. More conservative solutions have brought the number down to ten thousand possible civilizations in the Milky Way. Of course, if the rest of the universe is added to the mix, the number increases astronomically, to put it mildly. However, since most of the terms in the equation are a matter of conjecture—or faith, depending on your viewpoint—playing with the Drake Equation is still an intriguing piece of table talk but, ultimately, speculative. More than a few scientists have likened it to the old theological argument over how many angels can dance on the head of a pin.

For the sake of argument, accept that the ten thousand figure is a plausible estimate of the number of planetary civilizations out there trying to send us a message. If they are out there, how do we find them? Back in 1960, Frank Drake proposed that the best way to seek out intelligent life in the universe was to listen to the skies. He started Project OZMA, named after a character in the classic L. Frank Baum *Oz* series which, of course, included the *Wizard of Oz*. But his idea was no kiddy fantasy. Drake began listening for radio signals from space with the eighty-five-foot-wide telescope at Green Bank, West Virginia. Since then, the work has been concentrated in what is now widely known as SETI, the Search for Extraterrestrial Intelligence. Using radio telescopes, such as the one depicted in the film *Contact*, based on Carl Sagan's novel of the same name, at Arecibo, Puerto Rico, and elsewhere, astronomers from some of the leading institutions in the country work on SETI. Several years ago, NASA felt it was worth some funding, but Congress decided otherwise and withdrew financial support for SETI. The project is now privately funded, most notably at the University of California at Berkeley, where a SETI chair was endowed with $500,000 by two Berkeley alumni. These astronomers listen for any signals from the cosmos that appear to be transmissions from other planets. The search has also been extended to messages that may be found in flashes of light, perhaps from lasers.

The big difference in SETI now is that anyone can join in on the search and become the do-it-yourself version of the Jodi Foster character in the film version of *Contact*. You, too, can listen in on the search at the SETI website, if you are inclined. In addition, SETI created software in 1999 that home-computer users can download and use as a screen saver. With "SETI@home," anyone can analyze some of the billions of bits of data that SETI has collected but cannot process. By farming the material out to hundreds of thousands of home-computer users—by 2000, more than one million Web users had signed up to participate in SETI@home—the SETI group has created a league of thousands of amateur radio-astronomer extraterrestrial hunters. However, the extra help has not changed the box score to date. After more than forty years of searching, no bites yet.

Tossing a splash of cold water on the SETI concept is another group of serious, thoughtful scientists. They may be best represented by Donald C. Brownlee, the astronomer who heads NASA's $166 million Stardust mission to capture interplanetary dust, and Peter D. Ward, a geologist specializing in mass extinctions. In their 1999 book, *Rare Earth: Why Complex Life is Rare in the Universe*, these scientists speak for the "show me" camp when it comes to extraterrestrial life. They argued that despite all the numbers that "Saganists" can trot out to fill in the Drake equation, the question known as the "Fermi Paradox" still remains. As the physicist who helped construct the atomic bomb once asked his fellow bomb makers at Los Alamos, "Where are they?"

There are several possible answers. There is life elsewhere in the universe, but we haven't found it yet. Or it has not evolved as it did on Earth. In other words, there could be simple forms of life from amoeba to bacteria to slime molds that have not progressed past an early state and whose existence we simply haven't been able to find yet. Theoretically, there were, or could be, dinosaurs on another planet, just as there were on Earth once, for much longer than humans have been here. And they, too, have either been wiped out, as the dinosaurs were, or they have not evolved beyond dinosaurs, in which case, we won't be getting any radio calls from them. Or, finally, it may well be that Earth is the only planet that passes the Life Test. Rather than a "Mediocre Planet," as some astronomers would have it, Earth is a very elite place.

According to Brownlee and Ward, some of the key factors in making Earth the Garden of Eden planet are:

- The correct distance from the Sun, which moderates temperatures.

- The right mass of the Sun, which provides stable planetary orbits in which the other planets in the solar system do not create orbital chaos.

- The right planetary mass, which allows Earth to retain its atmosphere and oceans.

- Jupiter as a neighbor at the correct distance; the large planet acts as a buffer, clearing out comets and asteroids that might otherwise strike Earth with cataclysmic results.

- Plate tectonics that build up landmass and enhances biodiversity.

- Oceans in the right proportions.

- A large Moon at the correct distance to stabilize Earth's tilt.

- The correct tilt of Earth on its axis, which moderates seasons.

- Few giant catastrophic impacts in Earth's history. At least two, and possibly five, such impacts are thought to be responsible for mass extinctions that have occurred in Earth's history; one of these took place 250 million years ago, and a second, 65 million years ago, was responsible for the dinosaur extinction.

- The right amount of carbon; sufficient for life but not enough to create a runaway Greenhouse Effect, like on Venus.

- Biological evolution—the pathway to complex plants and animals.

- The correct position of the solar system in the galaxy—not in the center, where it would be subject to much greater radiation (*Rare Earth*, pp. xxvii–xxviii).

For these and a variety of other reasons, Ward and Brownlee believe Earth is unique. However, both scientists, who are professors at the University of Washington in Seattle, do not view their conclusion as a reason to shut down SETI, or the ongoing search for other signs of life elsewhere in the universe. On the contrary, they support scientific searches for signs of alien microbes and the radio hunt for alien civilizations. They also confess to an "Earth bias" and agree that life elsewhere could evolve in a form that is completely outside the Earth's experience.

Setting aside the speculative, what is the current best evidence for life elsewhere in the universe? Now come the tantalizing hints of possible life. First, the discovery of new planets (discussed in part III). Stars like the Sun have been found, with large planets the size of Jupiter. So far, there are no known Earth-like planets, but that is partly because we have not yet been able to get a close look at these very distant planets.

Then there is that controversial rock from Mars found in Antarctica with hints of fossilized bacteria. The jury is still out on that one. There is also the growing likelihood of liquid water once having been plentiful on Mars. Logic dictates that if Mars had water once, and Earth still does, then other planets could also have water, widely viewed as one of the key factors in permitting life to flourish. To bolster that idea, in February 2001, astronomers reported that orbiting observatories probing space around young and dying stars have found vast waves of water vapor and clear traces of carbon molecules that can play a basic role in organic chemistry—in other words, the basic chemistry set of life. Discovered in the dust and gas surrounding distant stars, these findings boost the theory that the "cosmic stew" of life exists elsewhere in the universe.

What are the ramifications of finding life elsewhere, as well as not finding it? The discovery of life elsewhere in the universe would be the story of the millennium, even if it is something as primitive as bacteria. Such a discovery would radically change the way we see ourselves in space. After all, life began a mere few billions of years ago on Earth, in puddles of warm water, as single-celled organisms that evolved into plants capable of photosynthesis that created oxygen.

But simmering beneath this scientific argument, there is also a theological debate. If Earth is unique, after all, doesn't that place our planet back on that pedestal it once occupied before Copernicus and Galileo came along and took us down a few notches? Even the Vatican now agrees with Copernicus and Galileo that the Earth is not the center of the universe. However, if it is a singular sensation—the only home to life in the entire vastness of the cosmos—then what does that uniqueness suggest? Was Earth just naturally selected to be the crown of creation? Or was there a design in that selection? Which leads, of course, to a bigger question. Who was the Designer?

PART V

THE OLD ONE'S SECRETS

The third angel blew his trumpet, and a great star fell from heaven, blazing like a torch, and it fell on a third of the rivers and on the springs of water. The name of the star is Wormwood. A third of the waters became wormwood, and many died from the water, because it was made bitter. The fourth angel blew his trumpet, and a third of the sun was struck, and a third of the moon and a third of the stars, so that a third of their light was darkened; a third of the day was kept from shining, and likewise the night. Then I looked, and I heard an eagle crying with a loud voice as it flew in midheaven, "Woe, woe, woe to the inhabitants of the earth." . . .

The Revelation of St. John the Divine 8:10–13

Quantum mechanics is very worthy of regard. But an inner voice tells me that this is not yet the right track. The theory yields much, but it hardly brings us closer to the Old One's secrets. I, in any case, am convinced that *He* does not play dice.

ALBERT EINSTEIN, in a 1926 letter

There is in space a small black hole
Through which, say our astronomers,
The whole damn thing, the universe,
Must one day fall. That will be all.

HOWARD NEMEROV, *Cosmic Comics*, 1975

When was Creation?

How did a clerk in the Swiss Patent Office change the world?

What became of Einstein's brain?

The Big Bang: Was it big and did it really bang?

"Red Shift, Blue Shift, One, Two, Three." Isn't that a Dr. Seuss book?

What did pigeon droppings have to do with the Big Bang?

What was there before the Big Bang?

Can anything else explain the creation of the universe, besides the Big Bang?

What are Inflationary theory, dark matter, and quintessence?

Open, closed, or flat: How will the universe end?

What Do GUTs, TOEs, and strings have to do with the universe?

"Yes, a hole in space three hundred million light-years across does make me pause and feel tiny and insignificant, but a glance around at my peers usually restores my equanimity."

It is one thing to ponder getting to the Moon and back. Or how to use the potty at zero-g. Or whether we should send people into space. Or whether we are alone in the universe. They are all reasonable, and practical questions. But thinking about space raises another set of questions entirely. Now muse over the Big Questions, the sublime enigmas of twentieth-century cosmology. These are the questions found in this closing section, which looks as far into the past as we can look and as far into the future as we can imagine. Thinking about space and the universe underwent a profound revolution during the twentieth century. To start with, this story moves backward, to the beginnings of time. This is about Big Ideas. The Beginning of the Universe. How it all began. Cosmology. Then comes a Really Big Question. It is significant, but, perhaps, less relevant than some of the other questions in this book, because most of us will not be around for the true "Final Answer." How will it all end?

The question of the fate of the universe is one of those speculative Big Questions that bring all the other questions in this book to the uneasy intersection of science and faith, knowledge and belief. It is the place where slide rules and calculators cannot go. It is there, perhaps, in "the high untrespassed sanctity of space," that we will get the ultimate answer. Do the scientists and philosophers who focus on the laws of Nature and dismiss a divine hand—the heirs of good old Thales of Miletus—have it right? Or will we continue to search and extend the human grasp, until humanity may someday reach the stars and touch "the face of God"?

VOICES OF THE UNIVERSE:
JAMES USSHER, *The Annals of the World* (1658)

According to our chronology, [the creation of the world] fell upon the entrance of the night preceding the twenty-third day of October in the year of the Julian calendar 710 [4004 B.C.].

When was Creation?

Bishop James Ussher of Armagh, Ireland, had a theory. Well, you could call it an idea, or even an inspiration. Through a careful reading of the Holy Bible, the infallible—and, in Ussher's view, unquestionable—divine Word of God, one could precisely reconstruct a chronology of the world. Using this method of calculating biblical ages as cited in all the books of the Bible going back to Genesis, Bishop Ussher worked backward through the Scriptures until he found his answer. According to Bishop Ussher, God created the world in the beginning of the night on the twenty-third day of October, 4004 B.C.

The good bishop was working in 1658, of course, and before you dismiss Ussher as being naive or blinded by his faith, consider that both Johannes Kepler and Isaac Newton—the great scientific minds of the Enlightenment—were both interested in calculating this date. Both men had arrived at dates close to Ussher's using similar methods. However, Bishop Ussher's calculations were accepted by much of the Western world for the next two and a half centuries—in fact, they were often printed in the margins of Bibles giving them the veneer of Holy Writ. With the advantage of two and a half centuries of science, many people now dismiss Ussher's chronology—although there are many others who still accept it—as an amusing mistake. But it wasn't so much that Bishop Ussher was foolish or blind. He was simply working with incomplete and, sometimes, erroneous data.

Then along came science. By the end of the nineteenth century, people like Charles Darwin upset the theological apple cart by suggesting a very different possibility for the development of life on Earth. His theory of "natural selection" has been amply supported during the past century. Even the state of Kansas, where religious fundamentalists on the state's school board voted to remove evolutionary theory from its schools a few years ago, relented in 2001. Kansas has placed evolution back in the science class, where it belongs. Along with Darwin came a generation of geologists and earth scientists who began to understand that Earth was a much older place than Bishop Ussher's chronology could possibly justify.

So by the turn of the twentieth century, the neat cosmology of

Bishop Ussher was turned on its head. Evolution, geological upheaval, "continental drift" (now usually called "global plate tectonics"), and the discovery of radiation were all large steps in rewriting the history and age of Earth. But they still left a mystery the bigger question of the birth and age of the universe.

Then, in 1905, a relatively (no pun intended) young and unknown German mathematician named Albert Einstein published his special theory of relativity, which showed that measurements of space and time vary according to the observer's motion, and that mass and energy are equivalent. A couple of years later, a Russian mathematician named Herman Minkowski, Einstein's former college professor who had once derided his student as a "lazy dog," formulated a view of the universe in which the traditional three dimensions of space were combined with the additional dimension of time. "The views of space and time, which I wish to lay before you, have sprung from the soil of experimental physics and therein lies their strength," Minkowski wrote in a 1908 paper. "They are radical. Henceforth, space by itself and time by itself are doomed to fade away into mere shadows, and only a kind of union of the two will preserve an independent reality."

Einstein picked up that ball and ran with it in 1916, in his general theory of relativity, in which he described gravity as a curvature in the geometry of space-time, produced by mass and energy. Einstein later produced another paper, "Cosmological Considerations on the General Theory of Relativity," which reflected on the effects of matter and energy on the actual geometry of the universe. His work became the cornerstone of modern cosmology, particularly when Sir Arthur Eddington, a British astronomer, was able to measure the bending of light during a solar eclipse in 1919 and confirm Einstein's predictions. Around the same time, Dutch astronomer Willem de Sitter (1872–1934) used Einstein's theory of relativity to describe an expanding universe. And, in 1922, a Russian mathematician and meteorologist Aleksandr Friedmann, suggested a universe expanding from a dense earlier state. Then, in 1927, a Belgian Jesuit priest who was also a mathematician—or should he be called a mathematician who was also a Jesuit priest?—named Georges Lemâitre suggested that not only was the universe expanding, but by going back in time—just as

Bishop Ussher had tried to do—one would discover "a day without yesterday."

Lemâitre theorized that everything had started from a single point, which he called the "primeval atom" or "cosmic egg," when space was infinitely curved and all matter and all energy had exploded and expanded outward. He called this eruption the "big noise." The idea was not taken seriously by many scientists, including more than a few who dismissed any idea coming from a priest—a Jesuit, no less, like the priests who had tried Galileo three hundred years earlier—especially since this priest's idea seemed to imply a precise moment of creation that sounded vaguely biblical.

To that point, most of these ideas were theoretical until 1929 and the landmark discovery by Edwin Hubble that distant galaxies were moving away from one another. There was only one possible conclusion to be drawn from Hubble's discovery—the universe was expanding.

According to the Big Bang theory, the universe began between 10 and 20 billion years ago. Recent developments in this cosmic dating-game seem to have narrowed that down to 12.5 billion years ago, with a fudge factor of two or three billion either way, which brings us to 10 to 15.5 billion years.

To offer up a ballpark figure for the age of the universe as between 10 and 15 billion years, brings to mind a famous line from American politics. It is attributed to Senator Everett M. Dirksen, who was talking about federal budget dollars, not cosmology: "A billion here, a billion there, and pretty soon you're talking about real money."

Well, a billion years here, a billion years there, and pretty soon you're talking about a real old universe.

VOICES OF THE UNIVERSE:
ALBERT EINSTEIN, at age sixteen

What would the world look like if I were sitting on a beam of light, moving at the speed of light?

How did a clerk in the Swiss Patent Office change the world?

The patent applications the young man had to deal with at the turn of the twentieth century look rather quaint, ironic, and pointless from the perspective of the turn of the millennium: one was for an improved pop gun, another for controlling alternating electrical currents. That particular application, the clerk commented, was "incorrect, inaccurate, and unclear." He was right about that one. You might say, when Albert Einstein spoke, people listened. Except that, in 1906, not many people did listen. And even if they had, they would have been hard pressed to grasp what this Patent Office clerk was saying.

Albert Einstein (1879–1955) is an instantly recognizable icon of modern times. His very name is now synonymous with genius, or lack of it—as in, "You're no Einstein." His image was used by Apple to help introduce its "Think different" advertising campaign for a new generation of Macintosh computers and by Pepsi in another series of ads. He has inspired bad movies—Walter Matthau portrayed him in a forgettable romantic comedy called *I.Q.*—and myths and legends about everything from his capabilities as a child, to his illegitimate daughter, to his brain and its whereabouts.

Born on March 14, 1879, in Ulm, Germany, Albert Einstein was slow to talk and not an especially good student in his early years. The idea that he suffered from dyslexia has never been confirmed; more likely, he was bored with the style of nineteenth-century schools and was probably far ahead of his classmates. At age five he was given a compass that fascinated him and a violin that instilled in him a passion for the mathematics in music. Two uncles were instrumental in fostering his love for math and physics. Surely he was not suited to the stiff, rigorous regimentation of German schools of the time. Expelled from one school at fifteen, after he was told that his presence in class was disruptive, Einstein would later say, "Humiliation and mental oppression by ignorant and selfish teachers wreak havoc in the youthful mind that can never be undone and often exert a baleful influence in later life."

Clearly, he was "thinking different," as Apple would have it, from an early age. He once told an interviewer in 1935, "As a boy of twelve years, making my acquaintance with elementary mathematics, I was thrilled in seeing that it was possible to find out the truth by reasoning alone, without the help of any outside experience. . . . I became more and more convinced that even nature could be understood as a relatively simple mathematical structure." Somewhere, old Pythagoras must have been smiling when he heard that.

After a failed attempt to enter Zurich's prestigious Polytechnic Academy when he was two years too young, Einstein was accepted to the school in 1896, the same year he relinquished his German citizenship because he disliked German militarism. While at the academy, he met a fellow student who would become his wife, Mileva. After school, he looked for a teaching position in the academic world, without success, and relied upon work as a private tutor and substitute teacher until a friend helped arrange a job with the Swiss Patent Office in Bern in 1902.

In 1903, Einstein married his first wife, Mileva, a few months after their daughter, Lieserl, was born out of wedlock. The fate of the child, who had scarlet fever and may have been put up for adoption, is a mystery. Einstein feared that if his superiors learned of an illegitimate child, his appointment might be jeopardized. She never lived with her parents, and Einstein never saw her or mentioned her again after a single reference in a letter. All traces of her were lost. (In a 1999 book *Einstein's Daughter: The Search for Lieserl*, Michele Zackheim concludes that Lieserl may have been severely mentally handicapped, perhaps suffering from Down syndrome, but indeed died of scarlet fever in 1903.)

While working in the Swiss Patent Office, Einstein rode a tram to work, examined patent applications, and, after finishing a day's work, headed for the Café Bollwerk where he met with some of the students he privately tutored in physics to supplement his clerk's salary. Then, in 1905, Einstein published a series of papers that changed science and mathematics as profoundly as Newton's *Principia* had. In "On the Electrodynamics of Moving Bodies," he answered his ten-year-old question about riding the beam of light. This theory was called the "special theory" of relativity and, essentially, showed that measurements of space and time vary according to the observer's motion. In other words, passengers who are traveling smoothly in a train cannot tell whether they are moving or not unless they look out the window. The situation becomes different if the train accelerates. Then the passengers will feel a slight push in the direction opposite to that in which the train is moving. In a related paper, he established the equivalence of mass and energy—which, in the original, was written: "If a body emits the energy L in the form of radiation, its mass decreases by L/V^2." In a 1912 manuscript, according to a collection of Einstein's papers, *The Expanded Quotable Einstein*, Einstein rewrote the equation in its more famous form: $E=mc^2$. By establishing the relationship between energy, mass, and speed of light, Einstein laid the theoretical foundations of the atomic bomb. The third of five papers he published that year concerned the photoelectric effect, for which he earned a Nobel Prize sixteen years later.

In 1906, Einstein received his doctorate from the University of Zurich and got a promotion as well. He was now a "Technical Expert, Second Class," probably the smartest second-class technical expert in the history of the world. It took another three years before he received a professorship, at which time he finally left the Patent Office.

In 1916, in the midst of World War I, Einstein published "The Origins of the General Theory of Relativity," in which he described gravity as a curvature in the geometry of space-time, produced by mass and energy. Einstein's prediction of the curvature or bending of light was first confirmed by English astronomer Arthur Eddington in 1919, during a total eclipse of the Sun. Working from Africa, Eddington photographed stars in a briefly darkened sky. The results of his expedition confirmed Einstein's theory—the light rays coming from the stars were found to be offset, just as Einstein had predicted. Much later, in 1976, the *Viking* space probes that reached Mars provided more precise confirmation of general relativity. The Sun also bends and delays radio waves, and this delay was measured by sending radio signals between the Earth and Mars. According to general relativity, massive bodies that orbit one another emit gravitational waves. Further observations from space have only served to strengthen the validity of Einstein's theories.

Einstein had, by that time, acquired worldwide notoriety. He said around 1920, "At present, every coachman and every waiter argues about whether or not the relativity theory is correct." At the same time, his success with relativity disguised his problems with his family. His marriage to Mileva produced two more children: Hans Albert

(1904–1973) and Eduard (1910–1965), who suffered from schizophrenia and died in a psychiatric hospital; after leaving Europe in 1933, Einstein had no contact with his second son, for reasons he later said he could not analyze. His marriage to Mileva had apparently been one of obligation, and they divorced in 1919. A few months later, Einstein was married to his cousin, Elsa Lowenthal, with whom he had begun an affair in 1912. But it later became known that Einstein had actually asked one of Elsa's two daughters, Ilse, to marry him, and was turned down, before he married her mother.

In 1933, after the Nazis came to power, Einstein left the country, and, eventually, came to America in the fall, to Princeton University in New Jersey, where he would spend the rest of his life. In 1939, he signed a famous letter to President Franklin D. Roosevelt, warning of the military implications of atomic energy. That letter set in motion the Manhattan Project to build the atomic bomb.

Einstein's work contained two other assumptions: the universe is *homogenous*—meaning, it is essentially the same everywhere—and *isotropic*—meaning, it is the same in all directions. When applied to cosmology, Einstein's work predicted that the universe must either expand or contract, a conclusion he did not particularly like. To adjust for this seeming contradiction, Einstein theorized something he called a "cosmological constant," a term that would allow the universe to be static, an idea that he later called the "biggest blunder of my life." In fact, recent discoveries suggest that it may not have been a blunder at all. Analysis of photographs of an exploding star made by the Hubble Space Telescope in 1997 confirmed Einstein's conjecture: that all of space is filled with an invisible form of energy that creates a mutual repulsion between objects normally attracted to each other by gravity. The confirmation of Einstein's theoretical "repulsive gravity" would, in the minds of many scientists, be worth another Nobel Prize (see question on "quintessence" on page 304).

VOICES OF THE UNIVERSE

ALBERT EINSTEIN on atomic weapons

I do not know how the Third World War will be fought, but I can tell you what they will use in the Fourth—rocks! (From a 1949 interview in *Liberal Judaism*)

I made one mistake in my life—when I signed that letter to President Roosevelt advocating that the bomb should be built. But perhaps I can be forgiven for that because we all felt that there was a high probability that the Germans were working on this problem and they might succeed and use the atomic bomb to become the master race (conversation recorded in the diary of Linus Pauling, *The Expanded Quotable Einstein*, p. 185).

What became of Einstein's brain?

Albert Einstein, "The Man of the Century,"according to *Time* magazine, died in Princeton Hospital on April 18, 1955. Both during his life and since his death, he has inspired anecdotes, myths, and phantom quotations meant to give authority to statements he supposedly made. For instance, Einstein never said, "We use only 10 percent of our brains." Nor did he once say that his "fondest dream was to become a geographer."

By the end of the war, Einstein joined other physicists in a disarmament movement. In 1947, Einstein, who became an outspoken supporter of world disarmament, told a *Newsweek* reporter: "Had I known that the Germans would not succeed in producing an atomic bomb, I never would have lifted a finger." He became a committed pacifist and called for a world government that would protect humanity from self-destruction. His death came after he wrote his last signed letter, a manifesto urging all nations to renounce nuclear weapons. In it, he wrote: "There lies before us, if we choose, continued progress in happiness, knowledge, and wisdom. Shall we instead choose death, because we cannot forget our quarrels? We appeal, as human beings, to human beings: remember your humanity and forget the rest."

Such sentiments, coming in the Cold War, brought Einstein under suspicion, and J. Edgar Hoover's FBI treated him as a potential subversive in the Cold War climate of fear in the 1950s. A 1,500-page dossier was compiled about Einstein, who had not been given security clearance to work on the Manhattan Project during the war.

Curiously, few aspects of his life and death have attracted as much attention as his brain. His brain and eyes were removed and preserved by two pathologists who performed an autopsy, most likely, against Einstein's own wishes. Fearful that his remains would become some sort of shrine, he had requested cremation, and his ashes were secretly thrown into the Delaware River. After learning about the brain's removal by Dr. Thomas Harvey, Einstein's family agreed that it could be used for scientific study but not commercial purposes. Dr. Harvey, who later kept the brain in glass jars at his home in Kansas, mailed sections of the brain, which is in the normal range in size and weight, to various researchers over the years. One of them, Professor Marian Diamond of Berkeley, California, reported that Einstein's brain contained an above-average number of glial cells, which nourish neurons in the brain's left hemisphere. And in June 1999, Canadian neuroscientist Sandra Witelson reported her discovery that the part of Einstein's brain thought to be related to mathematical reasoning was wider than normal and that the fissure, or groove, that runs from the front to the back of most brains did not extend all the way back in Einstein's case. Many other neurologists dispute these findings. Einstein's eyes, removed by opthalmologist Henry Abrams, remained in the doctor's possession.

VOICES OF THE UNIVERSE:
STEVEN WEINBERG, *The First Three Minutes* (1977)

In the beginning there was an explosion. . . . At about one-hundredth of a second [later] . . . the temperature of the universe was about one hundred thousand million degrees Centigrade. This is much hotter than in the center of even the hottest star, so hot, in fact, that none of the components of ordinary matter, molecules, or atoms, or even the nuclei of atoms, could have held together. Instead, the matter rushing apart in this explosion consisted of various types of the so-called elementary particles.

As the explosion continued, the temperature dropped . . . reaching thirty thousand million degrees Centigrade after about one-tenth of a second. . . . The energy released in this annihilation of matter temporarily slowed the rate at which the universe cooled, but the temperature continued to drop, finally reaching one thousand million degrees at the end of the first three minutes. It was then cool enough for the protons and neutrons to begin to form into complex nuclei, starting with the nucleus of heavy hydrogen. . . .

The Big Bang: Was it big and did it really bang?

Do you remember learning about the Holy Roman Empire back in school? You have probably forgotten some of the details about this Germanic empire in western and central Europe, that began in A.D. 962. But what you may recall is the fact that every teacher says the same thing about it: "The Holy Roman Empire. It was neither holy, nor Roman, nor an empire." Or, maybe, you remember *Saturday Night Live*'s Mike Meyers as the "Coffee Talk" lady, Linda Richman, who says: "The peanut. It is neither a pea nor a nut. Discuss."

The concept of the beginning of the universe called the "Big Bang" is often described as an explosion. Well, it's a bit like that Holy Roman Empire, or a peanut. It wasn't big. It wasn't noisy. And it wasn't an explosion, at least, not like a Fourth of July fireworks display. Discuss.

The birth of the idea of the birth of the universe in a big bang goes back to the beginning of the century, and Einstein's theoretical work. But the phrase "big bang" is only half that old. In the history of insults, few have had greater unintended results than this one. In the 1950s, British astronomer Fred Hoyle—that's Sir Fred to you—had a BBC radio show called "The Nature of the Universe." (In an era when American radio is dominated by the likes of Howard Stern and Rush Limbaugh, the idea of a physicist talking about the creation of the universe on the radio sounds too good to be true.) Hoyle was asked a question about a theory of the beginning of the universe then being bandied about the cosmological coffee klatches. Hoyle, a witty and entertaining fellow, dismissively labeled the notion that the universe began in an explosive event that created all of space and matter as the "Big Bang."

This theory, born earlier in the century through the work of Einstein, Sitter, Friedmann, and Lemâitre, found a champion in the Russian émigré George Gamow (pronounced "Gamov," 1904–1968), a former student of Aleksandr Friedmann. In the 1940s, Gamow, working with two younger colleagues, Ralph Alpher and Robert Herman, believed that if the young universe was small and dense it must have also been hot and then cooled. They even calculated the temperature to which the cosmic leftovers would have cooled. It was this idea that Hoyle dismissed as the "Big Bang."

In a nutshell, the Big Bang, or, as some cosmologists prefer to call it, "the Standard Model of Cosmology," goes something like this. About 15 billion years ago, the universe erupted from an enormous and still largely unexplained event—often referred to as a "singularity"—from which all of space and matter were created. That's why you can't say it was an explosion. *Nothing* can't explode. And at the instant of the Big Bang, there was a *nothing*. It's a little like a cosmic episode of *Seinfeld*, the show about nothing. It also didn't happen anywhere—that is, at a single location—but Everywhere. In other words, since space didn't exist, there was no place for the Big Bang to happen. Or as Gertrude Stein once famously remarked about Oakland, California, "There was no there there."

For most people who are used to seeing, hearing, touching, tasting, and smelling our world, this is a hard concept. But can you wrap your mind around that? Something from nothing. Everywhere from nowhere.

Within a fraction of a second of time—0.00001 of a second, to be more precise—all matter and energy came into being. Before that, everything was unified, theoretically speaking, into one cosmic "seed," or "egg," as the priest Lemâitre had called it. The temperature of the universe within millionths of a second after this event was hot. No, not just hot. *Reeeally* hot. *Really, really hot.* Like, hotter than Texas chili hot. Hotter than New York City in August hot. Numbers that don't mean much to most of us—like 10 trillion, trillion times hotter than the core of our Sun. According to *Scientific American Science Desk Reference*, "Scientists have calculated that one million-million-million-million-million-millionth of a second after the Big Bang, the universe was the size of a pea, and the temperature was 10 billion million million million°Centigrade/18 billion million million million°F.

One second after the Big Bang, the temperature was about 10 billion°C/18 billion°F."

In that instant of "cosmic expansion," all matter, energy, time, and space were created, but not as we know them today. As time passed, the universe expanded and began to cool. Immediately after the Big Bang, the universe consisted chiefly of strong radiation. This radiation formed a rapidly expanding region called the *primordial fireball*.

As the universe swelled and cooled, the first bits or primordial matter—called *quarks*—started to pull together into the most primitive and earliest forms of atoms. Like the radiation, the matter continued to decrease in density after the explosion. In time, the matter broke apart in huge clumps. They became the galaxies, stars, and eventually, planets, which began to emerge as these clumps of elements were pulled together by gravity. Part of at least one clump became a group of planets—the Solar System.

For hundreds of thousands of years, matter consisted of a seething mass of superheated subatomic particles, buffeted by high-energy radiation. Today's universe is cold and quiet by comparison, but, at its edge, astronomers can still detect the afterglow of the Big Bang—and, with it, the beginning of time.

The Big Bang theory, like all theories, can't be "proved"—although theories can be disproved. But it has been supported and amplified over the years by a tremendous amount of evidence that bolsters its reliability and widespread acceptance. Today it basically rests on a few essential facts that may be "everything you need to know" about the Big Bang:

1. The universe is expanding, which we know thanks to Mr. Edwin Hubble (see next question), a fact that has been amply supported since Hubble first observed it in 1929.

2. Cosmic microwave radiation or the cosmic microwave background (CMB) fills space—the leftovers of the denser, hotter beginnings.

3. That radiation has cooled to the exact temperature that astronomers calculated it should have.

4. There is an abundance of deuterium and helium in the universe, which makes sense if the universe was once much hotter.

"Red Shift, Blue Shift, One, Two, Three." Isn't that a Dr. Seuss book?

One of the most significant developments in astronomy came in the late nineteenth century, when astronomers and physicists began to develop a method to determine the movement of stars toward or away from Earth. English astronomer William Huggins (1824–1910) and French physicist Armand-Hippolyte-Louis Fizeau (1819–1896) discovered that colors in the light spectrum—the separation of light into red, orange, yellow, green, blue, and violet, when passed through a prism—behaved in a way that was similar to sound waves in what is known as the Doppler Effect. Named after the Austrian physicist Christian Doppler, who described the effect in 1842, the Doppler effect is the change in frequency of sound, light, or radio waves caused by the relative motion of the source of the waves and their observer. For example, the pitch (frequency) of a police car or ambulance siren seems higher as it approaches and lower after it passes and begins to move away. In reality, the actual pitch of the siren remains constant.

Huggins and Fizeau discovered that astronomers could study the speed of a star by measuring the apparent change in the frequency of its light waves. They estimated the distance and motion of a galaxy or star by measuring its *red shift*, an apparent lengthening of electromagnetic waves radiated by an object as it moves away from Earth. A red shift may be seen when light from a galaxy or a star is broken up into a band of colors called the *spectrum*. Lines of certain colors will be shifted toward the red end of the spectrum if the galaxy, or star, is receding or moving away from the Earth; they shift to the blue end of the spectrum if they are moving toward us. In 1869, Huggins determined that Sirius is moving away from the Earth at about twenty miles a second.

Having discovered, in 1924, that the galaxies are different, Edwin Hubble took this information about the shift in light and made his second remarkable discovery of twentieth-century astronomy. Hubble discovered that the light shift of receding galaxies increased in proportion to their distance from us. In other words, the universe is expanding, with the farthest stars moving faster. The galaxies are still moving away from one another, and the best current evidence indicates that they will move apart forever.

What did pigeon droppings have to do with the Big Bang?

In 1965, two scientists at Bell Labs discovered radiation coming from all over the sky. It was the afterglow of the Big Bang.

The cosmic radiation was born when the universe was about three hundred thousand years old, long before the formation of stars, when the cosmos was one thousand times hotter and one thousand times smaller than it is today. Until then, it was so hot and dense that radiation was "tightly glued to matter" and could not escape until the universe cooled down a bit. Once the radiation escaped, it raced along with the expanding universe.

In 1964, two radio astronomers working at Bell Labs (what later became AT&T and was then spun off as Lucent technologies) were working on an antenna for the new Telstar communications satellite system. But no matter where Arno Penzias and Robert Wilson pointed the horn-shaped antenna, it picked up a hiss. Attempting to eliminate the bothersome hiss, the two men shooed away some nesting pigeons and removed the accumulated pigeon poop.

But the hiss remained. What Penzias and Wilson discovered by accident, other scientists at nearby Princeton University were actively seeking out. Eventually, they traced the sound to its source: the hiss was the radiation left over from the cosmic fireball in which the universe was created. Penzias and Wilson received a Nobel Prize for this discovery. This "cosmic microwave background radiation" has cooled off in the 13 to 16 billion years since the Big Bang, but it still fills the heavens. When cosmologists first measured it in detail, they declared they were seeing "the handwriting of God." You can see God's handwriting at home: When you get "snow" on your television set, you are receiving background radiation.

In 1990, within the first nine minutes of its operation, the Cosmic Background Explorer (COBE), gathered more information about background radiation than had previously been discovered since 1965. In detail that astonished astronomers, COBE confirmed that the universe was radiating at exactly the temperature predicted by Gamow, Alpher, and Herman in the 1940s. It also bolstered what Wilson and Penzias had discovered, and showed that this background radiation existed throughout the universe uniformly in every direction.

One may say that time had a beginning at the Big Bang, in the sense that earlier times simply would not be defined. It should be emphasized that this beginning in time is very different from those that had been considered previously. In an unchanging universe, a beginning in time is something that has to be imposed by some being outside of the universe; there is no physical necessity for a beginning. One can imagine that God created the universe at the instant of the Big Bang, or even afterwards, in just such a way as to make it look as though there had been a Big Bang. But it would be meaningless to suppose that it was created before the Big Bang. An expanding universe does not preclude a creator, but it does place limits on when he might have carried out his job!

What was there before the Big Bang?

"I've got plenty of nothin' / And nothin's plenty for me," Ira Gershwin had Porgy sing in *Porgy and Bess.*

It is a pretty good answer to this question. There might have been something before the Big Bang, but we've got plenty of nothing when it comes to knowing what precisely that might have been. Scientists cannot explain with any certainty why the Big Bang happened, so it is pure speculation to consider what came "before." It might be intriguing to contemplate as you sit with some college buddies and ponder the universe, but for the present, this discussion is rather pointless—unless you actually enjoy arguing the finer points of cosmology and theology. Time—along with space, matter, and energy—was created in the Big Bang, so there was no "before." The Big Bang, theoretically, was the creation of something out of nothing. You can't ask what was before, because there is no evidence on which to carve out a theory. In other words, the Big Bang is widely accepted as accounting for everything we have been able to observe about the universe, but it doesn't explain itself. We are clueless as to the source of the Big Bang.

Can anything else explain the creation of the universe, besides the Big Bang?

Apart from the Book of Genesis, as well as the Creation myths of almost every other society and culture that has ever existed, most scientists would say "No." That hasn't always been the case. During the twentieth century, there have been quite a few alternative cosmological models, most of which have dropped by the wayside, while the Big Bang has been solidified and buttressed during the past forty years by substantial evidence.

For a long time, steady-state theory was the prime contender. Devised in 1948 by three colleagues at Cambridge, England, the Big Bang opponents Fred Hoyle, Hermann Bondi, and Thomas Gold, steady-state theory maintained that the universe has always existed in its present state. As the galaxies move apart, the new matter that continuously is being created forms into new galaxies that replace those which have receded to infinite distances. This theory never explained where this matter comes from. Modern astronomical observations, especially the information about cosmic background radiation, have more or less consigned "steady state" to a state of extinction.

VOICES OF THE UNIVERSE:
VERA RUBIN on "Dark Matter in the Universe"
(*The Scientific American Book of the Cosmos*)

New tools, no less than new ways of thinking, give us insight into the structure of the heavens. Less than 400 years ago, Galileo put a small lens at one end of a cardboard tube, and a big brain at the other end. In so doing, he learned that the faint stripe across the sky, called the Milky Way, in fact comprised billions of single stars and stellar clusters. Suddenly, a human being understood what a galaxy is. Perhaps in the coming century, another—as yet unborn—big brain

> will put her eye to a clever new instrument and definitively
> answer, What is dark matter?

In 1965, astronomer Vera Rubin became the first woman permitted to observe at Palomar Observatory. A member of the staff at the Carnegie Institution in Washington, D.C., she is one of the most distinguished scientists in recent history and was awarded the National Medal for Science by President Clinton in 1993. She is largely credited with shaping the theory behind "dark matter."

What are Inflationary theory, dark matter, and quintessence?

Although for most of the 1990s, Alan Greenspan, chairman of the Federal Reserve, came to be seen as the Master of the Universe, there are other forces at work, including a new strand of thought called *inflationary theory*, which has nothing to do with the Consumer Price Index and the Federal Reserve Board. A recent supplement to Big Bang thinking, inflationary theory basically states that, in its very earliest stages, the universe expanded at a much faster rate than it is expanding today. Based on the work of Dr. Alan Guth in 1980, and expanded upon by cosmologists such as Dr. Andrei Linde of Stanford University, inflation doesn't replace the Big Bang as much as it adds a layer of complexity to it. It attempts to explain why regions of the universe, even though they are separated by vast distances, still appear similar. Inflation suggests that the infant universe, the tiny speck of "primordial nothingness," which was somehow filled with intense energy, doubled in size *exponentially* in the tiniest fraction of a second after the birth of the universe. After this initial short period of inflation, the universe changed its rate of expansion to a *linear* expansion. The difference between exponential and linear can be represented by numbers: exponential means doubling with each step (1, 2, 4, 8, 16); a linear progression takes one step at a time (1, 2, 3, 4, 5 . . .)

One intriguing possibility of inflationary theory is the potential that inflation could have created universes beyond the scope of our own. Cosmologists are only now beginning to consider the possibility of *multiverses.* And you probably were thinking by now that one universe was already too much to contemplate and understand.

Along with accepting the Big Bang—with or without the inflationary add-on, which has gained favor in recent years—most scientists have also come to accept the existence of so-called *dark matter.* Although it sounds like the stuff of a *Star Trek* episode, this strange element was thought to account for as much as 90 percent of the universe. Since the 1930s, astronomers had begun to realize that there was matter in the universe that couldn't be seen—or detected—but it must be there. Why did they think that? Because something was providing the gravity that kept everything as it should be. The major discovery was made by a student of George Gamow, Vera Rubin, who had been told by Princeton University in the 1950s that it was biologically impossible for her to get her doctorate there—they did not accept women at the time. Rubin found that outlying stars in galaxies orbited as rapidly as inner stars. Something had to account for the speed of outer stars and for keeping them in their orbits.

The evidence that most of the mass of the universe consists of dark matter clumped around the galaxies is solid. But there is less agreement over what dark matter is. Two camps have developed recently, and, just to prove that astronomers have a sense of humor, they call their competing types of dark matter MACHOs and WIMPs.

MACHO stands for "Massive Compact Halo Objects" and may include burnt-out stars, white dwarfs, brown dwarfs, and dust. In 1996, astronomers announced findings that half of the dark matter in the Milky Way is made of MACHOs.

WIMPs, or Weakly Interacting Massive Particles, is a class of yet-to-be detected very small things that go whizzing by in the night.

If inflation, WIMPs, and MACHOs don't have you hankering for a nice comfortable chair in which you can rest all of your matter for a while, just wait. A major new development in cosmology may complicate the issue even more. Since the mid-1990s, some cosmologists have begun to suggest that even the known elements in the universe and the dark matter *still don't add up,* and account for less than half the contents of the universe. The latest theory is that this "funny energy," which is also called "dark energy," may actually account for more than half of the universe. The trick about this highly theoretical "dark energy" is that it is repellent. Literally. Unlike everything else in the universe, which exerts gravitational pull and attracts, "dark energy" repels things. Gravity pulls the chemical elements and other matter into stars and galaxies, but dark energy may be pushing back. It is this repulsive force, this theory suggests, that causes the universe to accelerate, a conclusion that has received some qualified support.

Recently, cosmologists have begun to call this dark energy *quintessence*. *Quintessence* means "fifth element," a throwback to ancient Greek cosmology of Aristotle's time, which suggested that Earth was composed of the four elements of earth, air, fire, and water, but that a *fifth element*—a solid crystalline substance—held up the Sun, Moon, and planets.

Most scientists expect that many more answers about inflation, dark matter, and quintessence will be answered with the Microwave Anisotropy Probe (MAP), which is designed to provide the most comprehensive picture yet of the entire sky. This $145 million probe, planned for launch in June 2001, will be placed in an orbit much farther from Earth than its predecessor, the Cosmic Background Explorer (COBE). This satellite will collect new and more detailed information about the cosmic background radiation, which is the best evidence for what the early universe was like. It should produce information on the various kinds of matter in the universe—both the ordinary stuff of which stars and people are made, as well as dark matter and, perhaps, even quintessence.

Open, closed, or flat: How will the universe end?

Science tries to ask "How?," and each answer is a victory. But asking how something *did* happen is very different from asking how it *will* happen. How the Universe will end is one of those questions.

Experts differ on whether the universe will continue to expand indefinitely. There are three generally accepted possibilities. For many years, astronomers presumed that all the galactic clusters that have been rushing apart since the Big Bang appeared to be slowing down. However, recent evidence seems to contradict that presumption. Nevertheless, the Big Bang's explosive moment of creation might have catapulted the material of the universe with sufficient energy to escape from the gravitational pull of the rest of the material. It would be like a rocket going fast enough to escape Earth's gravity. If that escape velocity were reached, the universe could conceivably expand forever. Astronomers and cosmologists call that an *open universe*.

On the other hand, if the material is not traveling with sufficient energy to continue infinitely, the universe will eventually stop expanding and fall back in on itself. That is called a *closed universe*.

A third possibility is that the universe will continue to expand, but the rate of expansion will slow over time in a way that will keep it expanding infinitely. Sometimes called the *marginally open universe*, this theoretical construct is also described as the *flat universe*.

If the first or third scenario is in our universal future, get out the thermals and start singing that old country favorite, "Turn out the lights, the party's over." An open, or flat, universe will continue to expand until the last star has burned out. All that would be left is a large number of burnt-out lumps of dead stars, separated by phenomenally large expanses of space. It will be dead and cold, and this fate of the universe is known as *the Big Chill*.

On the other hand, if the second scenario is correct—and the universe is closed, then it will fall back in on itself someday in what is called *the Big Crunch*. One possible outcome of that scenario is that by collapsing back down to an infinitely small, dense spot, the Crunch could lead to another Big Bang and begin the whole process all over again.

Astronomers and cosmologists hope and expect that the planned expeditions of the Next Generation Space Telescope and other space-based observatories may provide some of these answers. One of these, the $145-million Microwave Anisotropy Probe, or MAP, will open the whole of the universe, to be charted as the explorers of centuries past once mapped the seemingly unchartable oceans. Its observations could provide the key to the question of how fast the universe is expanding and, perhaps, hint at where it will all end.

VOICES OF THE UNIVERSE:
STEPHEN HAWKING, A *Brief History of Time* (1988)

> If we do discover a complete [unified] theory [of the universe], it should be in time understandable in broad principle by everyone, not just a few scientists. Then we shall all, philosophers, scientists, and just ordinary people, be able to take part in the discussion of the question of why it is that we and the universe exist. If we find the answer to that, it would be the ultimate triumph of human reason—for then we should know the mind of God.

What Do GUTs, TOEs, and strings have to do with the universe?

The brilliant physicist Stephen Hawking—who sits in the "chair" once occupied by Isaac Newton at the University of Cambridge—is discussing a theory that explains it all—a Theory of Everything, or TOE, an explanation for the fact that some laws of the universe conflict with others. This idea, which Hawking has poetically christened "the mind of God," has been given a much less poetic name: GUT—for Grand Unified Theory. The physicists who are looking for a GUT are trying to bring it together, in mathematical terms, the very forces of nature—the gravitational force that holds planets and stars together; the electromagnetic force that holds atoms together; the weak atomic force that causes the slow decay of particles known as radioactive decay; and the strong atomic force that holds together the atomic nucleus. Ultimately, they hope to show that each force is the same thing occurring in different ways.

In one sense, you can say we are halfway there. In 1979, the Nobel Prize in Physics was awarded to Steven Weinberg, Abdus Salam, and Sheldon Glashow for their work in combining the electromagnetic and weak interactions into what is called the "electroweak force."

In the past few years, cosmologists and physicists have begun to focus on a leading candidate for a GUT, known as "string theory." Very simply put, it attempts to describe the universe and all nature as composed of tiny, one-dimensional wiggling strings that vibrate in many dimensions—our familiar three dimensions, plus the fourth dimension of time-space. Developed in the 1980s, string theory describes elementary particles that are smaller than particles such as a proton. As Brian Greene described it in his bestselling description of string theory, *The Elegant Universe*: "According to string theory, if we could examine these particles with greater precision—a precision many orders of magnitude beyond our present technological capacity—we would find that each is not pointlike, but instead consists of a one-dimensional *loop*. Like an infinitely thin rubber band, each particle contains a vibrating, oscillating, dancing filament that physicists . . . have named a *string*." Green continues, "Just as the strings on a violin or on a piano have resonant frequencies at which they prefer to vibrate . . . the same holds for the loops of string theory. . . .

In principle, string theory can explain all the forces of nature. But even proponents of string theory acknowledge that their equations are just approximations—although *really* good guesses. Others dismiss "stringy physics"—one can almost hear the same snickering derision that Fred Hoyle may have used when he first said "Big Bang"—as untestable.

What does all of these arguing over inscrutable "strings" and "GUTs" have to do with the rest of us? Well as Brian Greene succinctly answers, "The discovery of the T.O.E.—the ultimate explanation of the universe at its most microscopic level, a theory that does not rely on any deeper explanation—would provide the firmest foundation on which to *build* our understanding of the world."

Beginnings and endings. Cosmic questions. Challenging concepts that press our very ideas of what is real or possible. This book concludes where Stephen Hawking and modern cosmology leave us, at the uneasy intersection of science and faith, knowledge and belief. For centuries, these opposing "faiths"—science and religion—have battled each other for supremacy. Cosmology as the Greeks understood it—order of the universe—has come a long way, but modern cosmology cannot really answer the question of "God." (You may, of course, freely substitute any of the many alternative names for the supreme deity.) That hasn't stopped people from trying. In traditional

Western philosophy, the answers have fallen, for centuries, into a few broad categories:

- **The Cosmological Argument:** This idea dates to Aristotle and holds that you can trace cause and effect only so far. Eventually, it has to start somewhere, with a First Cause, or Aristotle's Unmoved Mover. Philosophers have picked this one apart since the heyday of the Enlightenment, starting with the idea that there is no reason to believe that there aren't an infinite series of causes.

- **The Ontological Argument:** This vein of thought originated with St. Anselm and, to simplify it, goes something like this: We recognize that there is perfection. God is perfect. He must exist, since, if he didn't exist, he wouldn't be perfect. If that sounds like a dog chasing its own tail, that is how many philosophers view it.

- **Argument from design (a.k.a. the Teleological Argument)** Once upon a time, this was the cosmic-clockmaker theory. If you found a perfectly designed pocket watch in the woods, you would have to assume that it was designed for a purpose. This argument, basically, holds that the universe is perfectly designed, so something or someone had to work out all the details. That does not imply that this Clockmaker is still involved with the day-to-day function of the clock. "He" made it but doesn't have to wind it to keep it going.

All of these arguments are also met by "godless" notions of modern cosmology that include such ideas as random chance and chaos. As Timothy Ferris succinctly put it in a wonderful summary of cosmology through the centuries, *The Whole Shebang*, "If the world emerged from chaos and works by chance, what role can there be for an omniscient creator?" (p. 309).

Just to complicate matters, however, Ferris then rebuts his own question: "To find evidence of randomness in nature does not prove that there is no God. . . . First it is impossible to prove conclusively that what appears to be random really *is* random. . . . And even if the universe did arise from chaos, a believer could reasonably argue that God elected chaos as best suited for that purpose" (p. 310).

Ferris's arguments and counterarguments bring to mind another set of arguments, which first raised many of these questions in a very different context. The most skeptical, in many ways cynical, book ever written is an ancient Semitic folktale, known to many people as the Book of Job. Composed around the late sixth century B.C., perhaps when the Jews were in exile in Babylon, the biblical Job may be based upon a much older, ancient Near Eastern folk tale that asks the very modern question: "Why do bad things happen to good people?"

Job, a happy, wealthy, righteous man with a large family and flocks, is visited with plagues, pestilence, and loss when Satan challenges God to take everything from this man. Satan thinks that Job will curse God if all his worldly goods are destroyed. Without warning, everything, including his children, is taken away from Job. Egged on by a few friends, Job begins to question God, because he thought he had done everything right.

Out of the whirlwind, a none-too-happy God shows up to answer Job. In one passage, the Lord strikes a particularly astronomical note as he berates Job for even daring to question Him:

> "Can you bind the chains of the Pleiades,
> or loose the cords of Orion?
> Can you lead forth the Mazzaroth in their season,
> or can you guide the Bear with its children?
> Do you know the ordinances of the heavens?
> can you establish their rule on the earth?"

Job, humbled, and repentant, says to God,

> "I know that you can do all things,
> and that no purpose of yours can be thwarted. . . .
> Therefore I have uttered what I did not understand,
> things too wonderful for me
> which I did not know."

Clearly we haven't learned Job's lesson. Nor the lessons of the Tower of Babel. Nor those of Icarus and Wan Hu and Giordano Bruno. We are the children of Job, still daring to question. In the

Book of Job, God also asked, "Where were you when I laid the foundation of the earth? . . . Who determined its measurements—surely you know."

Not yet. But we can't seem to stop trying. Remember, after all, that other old story from the Bible—the one about the forbidden fruit? It was the fruit of the tree of knowledge. Adam and Eve ate from it, and we still haven't stopped looking for answers.

AFTERWORD

WHAT COLOR IS
YOUR UNIVERSE?

Kermit the Frog would have been pleased. In January 2002, two researchers at Johns Hopkins University announced that the universe was green. Working with data on star formation gathered from the Hubble Space Telescope and other deep-space probes, these astronomers analyzed the information with a device called a spectroscope and concluded that the predominant color of the universe, based on its chemical composition, was green. Actually a bluish-green. Call it turquoise. This announcement was meant to be a light-hearted sidebar to the scientists' more serious research.

But there was a computer glitch. A faulty software program had been discovered. A few months later, in March 2002, the same scientists issued a somewhat red-faced correction. The data actually revealed that the universe's predominant color was . . . *beige*!

In my original introduction to this book, I pointed out that what we know about space changes all the time. And since this book was completed and first published in 2001, the remarkable pace of exploration, innovation, research, and discovery in space has continued. So, what follows are highlights of some of the remarkable discoveries and landmarks that have taken place since this book's original publication.

For instance, now that we think we know what color the universe is, we also know a bit more about how big it is. In March 2002, Space.com reported that astronomers in Hawaii had spotted a galaxy that is thought to be 15.5 billion light-years away, the most distant object observed to date. We also know now that there are more than one-hundred "extra-solar" planets, or planets circling stars outside our solar system. (For more on the size and age of the universe, see pages 87, 289–290.)

Within our solar system, the discoveries also mount. We have landed a spacecraft on an asteroid and had another craft follow a comet. In July 2001, astronomers announced the discovery of twelve previously unseen satellites, or moons, orbiting Saturn. Using telescopes equipped with sensitive light detectors, they upped Saturn's moon count to thirty. Then, in May 2002, astronomers in Hawaii also announced the discovery of eleven new moons orbiting Jupiter, bringing the number orbiting the solar system's largest planet to thirty-nine, making it the "mooniest" planet. So far, that is. More moons are certain to be found in our solar system, as future missions scour the solar system more thoroughly. For instance, very little of the area around Jupiter, the largest planet, has yet been searched.

That'll be $238 billion. You want fries with that?

The devastating and tragic events of September 11, 2001, altered our lives in ways large and small. One immediate change was the transformation of America's budgetary priorities. Defense spending, already strongly supported by the Bush administration, moved to the head of the class. Supporters of the controversial Missile Defense System, once known as "Star Wars," pointed to the terror threat as justification for ramping up the development of a viable missile defense that could knock out an incoming enemy missile. Even though the program's critics pointed out that small-scale terrorist operations or biological and chemical attacks posed a far greater risk to American security than another nation's missiles, research on the Strategic Defense Initiative (SDI) was given new urgency.

In March 2002, the Pentagon announced that a test of a rudimentary interceptor was successful and a target missile had been destroyed. However, critics noted that the "kill vehicle" knew what the target looked like, where it was going, and when it would be launched. At about the same time, a congressional inquiry reported widespread technical failures and possible fraud in the testing of some components of the missile system. The debate over the development and deployment of the system continues.

Whichever point of view prevails, the nonpartisan Congressional Budget Office, early in 2002, put a price tag on the Bush administration's program for stopping a missile attack: $238 billion by the year 2025. (For more on "Star Wars," see pages xxii, 84–85.)

Water on Mars?

The answer is a more definite "Yes." But bring skates, not swimsuits.

Mars had become something of an unmentionable around NASA since the loss of two spacecraft in 1999, a devastating fiasco for the space agency. But the 2001 *Mars Odyssey* gave NASA reason to cheer. Launched in April 2001, *Odyssey* reached Mars on October 24, 2001, and settled into an orbit above the Red Planet. Unlike the ill-fated 1999 landers that crashed or disappeared, *Odyssey* is an orbiter that will survey the planet from a distance of 249 miles (400 kilometers) and map out its composition. It has already paid dramatic dividends by showing that Mars holds vast stores of frozen water. Signs of liquid water are thought to be a major prerequisite for life on another planet.

Shuttle surgery

Another stunning feather in NASA's cap was the successful repair mission performed by the shuttle *Columbia's* astronauts in March 2002. After capturing the four-story Hubble telescope in an operation that involved record-setting space walks, shuttle astronauts performed nearly thirty-six hours of delicate repairs and upgrades to the ten-year-old space telescope, a procedure likened to doing open-heart surgery in outer space. New solar wings will provide added power to keep Hubble going through 2010, along with new equipment that includes an updated camera that will allow the observatory to see farther into the universe than ever before. (For more on Hubble's history, see pages 262–263.) Within weeks of the successful operation, Hubble was sending back dazzling new photographs of deep space, including visions of thousands of distant, previously invisible galaxies.

Was it really an asteroid that did in Barney and his pals?

When a scientist first suggested that an asteroid impact was responsible for the mass extinction of the dinosaurs, the idea was impolitely dismissed. However, decades of research seem to have bolstered the notion, and many scientists now accept that catastrophic impacts with an object from space were responsible for creating a cloud of dust that blocked out sunlight, killing much of life on Earth. Such an impact is thought to have not only caused the dinosaurs' demise but as many as four other mass extinctions in Earth's nearly five-billion-year history.

Another new theory also adds to, rather than contradicts, the asteroid impact theory. A group of scientists believe that massive climate change, possibly resulting from drastic changes in worldwide sea levels, might have wreaked havoc on the environment before the asteroid hit. In *Discover* (June 2002), leading "dinosaur hunter" Jack Horner speculates that the asteroid impact was the "knockout blow in a fight the dinosaurs already lost."

But there are dissenters. Some scientists dismiss the asteroid theory entirely. Others think it just needs some fine-tuning. According to *Discover* (May 2002), geologist Kevin Pope has cast doubt on the theory by arguing that even a very large asteroid could not create a globe-encircling shroud of dust that would choke life from the planet. But instead of dismissing the asteroid theory entirely, Pope claims that the asteroid impact created not a dust cloud but a thick cloud of sulfuric acid capable of blocking sunlight. That was the real culprit in the dino-doom.

Close encounters of a scary kind

While the killer asteroid theory has its critics, the threat of an asteroid impact has always attracted plenty of attention from the public and from Hollywood. The seriousness of that potential threat was underscored several times recently. In July 2001, an early evening fireball was seen in eastern parts of North America, with motorists and pilots alike calling authorities from Toronto down to Virginia. Astronomers attributed this dramatic daylight-hours light show to an exploding asteroid. The pieces of the asteroid burned up as they entered Earth's atmosphere, creating the brilliant flashes called "fireballs."

Then, in March 2002, an asteroid about 165 feet across, or large enough to demolish a large city, passed within 288,000 miles of Earth. Astronomers didn't notice this "near miss" until the asteroid was in our "rearview mirror." This asteroid had come from the direction of the Sun, and in the astronomical equivalent of a baseball player losing a fly ball in the glare, the asteroid wasn't seen until it had already passed Earth.

A few weeks later, news of an even larger asteroid, known as 1950 DA, caused new concern as astronomers announced that this large asteroid had a 1 in 300 chance of smashing into Earth. The good news is that we have plenty of time to figure out what to do if this asteroid remains on an earthbound collision course: the 1950 DA doesn't reach Earth for another 800 years. In other words, don't take out any asteroid insurance just yet. (For more on asteroid impacts, see pages 132–137.)

Who gets to go to the space station?

There is no longer any joking about unhappy and misbehaving passengers on airliners since security measures in airports have been overhauled. But now that space tourism is becoming a reality, what about unruly passengers in space? After the successful flight of Dennis Tito, a wealthy American, and Mark Shuttleworth, a South African millionnaire, who each paid the Russians a reported $20 million for the chance to fly to the International Space Station, rules for selecting the next travelers became a necessity. Early in 2002, NASA and its partners on the International Space Station agreed on a set of guidelines for future travelers to the orbiting outpost.

So if you want to book a slot, here is what you need to know: no delinquency or misconduct in prior employment; no criminal, dishonest, infamous, or notoriously disgraceful conduct (well, that rules out most politicians, athletes, corporate executives, and accountants); no habitual use of alcohol to excess; no abuse of narcotics, drugs, or other substances; no membership or sponsorship in organizations that adversely affect the confidence of the public. Checking off that set of requirements narrows down the list of potential space travelers to a group of church ladies in Des Moines. So far, they have expressed no interest in going.

Do you have to wash your hands in space?

If your mother always warned you about the germs on your hands, imagine what she would say if they put her on a space station? Current research has shown that bacteria production increases by as much as 200 percent in space.

Who wants all that bacteria? New generations of antibiotics will come from research on bacteria, and the future of antibiotic research and production may come from bacteria cultures in space. Antibiotic production on Earth is a $20 billion industry. Drug companies sponsoring the research are wondering whether it can be done more efficiently and more profitably in space? (For more on the International Space Station and commercial development in space, see pages to xxi–xxii, 264–265.)

The tragic shock of September 11 obviously brought home a grim reality. The combination of technology and the human imagination can be deadly and horrific. History has all too often shown that, when put to destructive uses by terrorists or governments, our inventions can have awesome, and awful, potential. In many ways, exploring space represents the opposite end of the spectrum—the best of what can be accomplished by combining technical prowess with the human imagination. What I have tried to illustrate and illuminate in this book is the long-standing human fascination with space and how that fascination has translated into progress. Moon landings, shuttles, and deep-space probes have transformed our ideas of space as well as life on Earth. When *Sputnik*, the first spacecraft, was launched forty-five years ago, few people could have envisioned a day when American and Russian men and women would live and work together on a space station. But that is the reality. Once enemies, now colleagues and friends. Surely that is reason for hope and one of the messages of this book.

Space to Earth: "We can do it again."

REFERENCES/RESOURCES

The rapid developments in the fields of astronomy and cosmology, along with the massive amounts of information coming from the Hubble Space Telescope and other deep-space probes, mean that much of the literature published before 1995 has become dated. In the field of astronomy, the following list reflects the most recently published works, along with some popular, distinguished classics, which must be read with the proviso that some information may no longer be accurate.

In addition to the books listed, in which the emphasis is on works for the general reader, this resource guide also includes periodicals and Web sites. There are many good general-interest magazines, which keep track of major developments in astronomy and related fields.

The greatest change during the past decade has been the explosion of the Internet universe. Information on Web sites can change daily, and the sites listed are reputable resources, usually affiliated with astronomical groups. Most have excellent links to other sites. But when using the World Wide Web, it is always important to check the source of information. "Johnny's Neato Astronomy Home Page" is not the equivalent of the NASA home page. The sites listed are a fraction of the many sites devoted to astronomy on the Web. These sites are active as of this writing, but many sites are subject to abrupt and sometimes unexplained changes of name or address. Web sites are also prone to include dated material. When using the Web, wise researchers always check when a page was last updated.

BOOKS

Andreadis, Athena. *To Seek out New Life: The Biology of Star Trek.* New York: Crown, 1998. Taking off from the popular television series, this book examines the scientific possibilities of the show.

Asimov, Isaac. *Isaac Asimov's Guide to Earth and Space.* New York: Ballantine Books, 1991. A quick question-and-answer guide to some basics of space by one of the most prolific science popularizers.

Baker, David. *Scientific American Inventions from Outer Space: Everyday Uses for NASA Technology.* New York: Random House, 2000. An interesting collection of thousands of products with origins in the space program, which demonstrates the value of the space program.

Barnes-Svarney, Patricia, ed. *The New York Public Library Science Desk Reference.* New York: Macmillan, 1995.

Berlinski, David. *Newton's Gift: How Sir Isaac Newton Unlocked the System of the World.* New York: Free Press, 2000. An in-depth look at the contributions of one the most significant scientific thinkers in history.

Bodanis, David. *E=mc^2: A Biography of the World's Most Famous Equation.* New York: Walker & Co., 2000. A lively account tracing, for the layperson, the background and development of Einstein's famous but largely misunderstood equation.

Bolles, Edmund Blair. *Galileo's Commandment: 2,500 Years of Great*

Science Writing. New York: Freeman, 1997. A collection of essays and excerpts by major scientific writers, from ancients, like Herodotus, to moderns, such as Stephen Jay Gould.

Boorstin, Daniel J. *The Discoverers: A History of Man's Search to Know His World and Himself.* New York: Random House, 1983. An exhaustive but fascinating study of the major leaps in civilization.

Bova, Ben, and Byron Preiss, eds. *Are We Alone in the Universe: The Search for Alien Contact in the New Millennium.* New York: ibooks, 1999. A collection of essays by a variety of writers on the subject of alien life and the practical search for it.

Brian, Denis. *Einstein: A Life.* New York: John Wiley, 1996. A thorough and comprehensive recent biography.

Brody, David Eliot and Arnold R. Brody. *The Science Class You Wish You Had: The Seven Greatest Scientific Discoveries in History and the People Who Made Them.* New York: Perigee Books, 1996. A highly readable approach to the discoveries of Galileo, Newton, and Einstein, and other key scientific developments.

Bronowski, J. *The Ascent of Man.* Boston: Little, Brown, 1973. A classic that examines the rise of civilization and scientific thinking, tracing developments over long periods of time.

Burrows, William E. *This New Ocean: The Story of the First Space Age.* New York: Random House, 1998. A finalist for the 1999 Pulitzer Prize, an exhaustive and comprehensive account of the history of the effort to explore space.

Columbia Encyclopedia, 5th edition. New York: Columbia University Press, 1993.

Crowe, Michael J. *Modern Theories of the Universe.* New York: Dover, 1994. A scholarly work that details recent physics theories for readers who want a more complex approach to modern physics.

Crowe, Michael J. *Theories of the World from Antiquity to the Copernican Revolution.* New York: Dover, 1990. A scholarly work that traces ancient theories.

Davis, Kenneth C. *Don't Know Much About® Geography.* New York: Morrow, 1992. An introduction to geography, with an emphasis on how the history of people's views of the world have changed.

———. *Don't Know Much About® the Bible.* New York: Morrow, 1998. An introduction to biblical history for the general reader.

Davis, T. Neil. *The Aurora Watcher's Handbook.* University of Alaska Press, 1992. A basic guide to the Northern Lights.

De Pree, Christopher, and Alan Axelrod. *The Complete Idiot's Guide® to Astronomy.* Indianapolis: Alpha Books, 1999. Basic introduction with an emphasis on skywatching.

Engelbert, Phillis, and Diane L. Dupuis. *The Handy Space Answer Book®.* Detroit: Visible Ink, 1998. A quick question-and-answer guide to basic astronomy.

Ferguson, Kitty. *Measuring the Universe: Our Historic Quest to Chart the Horizons of Space and Time.* New York: Walker & Co., 1999. A very readable overview of the advances made in astronomy through history.

Ferris, Timothy. *Coming of Age in the Milky Way.* New York: Morrow, 1988. A comprehensive but engaging history of cosmology.

Ferris, Timothy. *The Whole Shebang: A State of the Universe(s) Report.* New York: Touchstone, 1997. A challenging bestseller about cosmology.

Gallant, Roy A. *National Geographic Picture Atlas of the Universe.* Washington, D.C.: National Geographic Society, 1994.

Gleick, James. *Genius: The Life and Science of Richard Feynman.* New York: Random House, 1992. A best-selling biography of the noted American physicist Richard Feynman.

Gleiser, Marcelo. *The Dancing Universe: From Creation Myths to the Big Bang.* New York: Dutton, 1997. A challenging overview of cosmological theories.

Greene, Brian. *The Elegant Universe: Superstrings, Hidden Dimensions, and the Quest for the Ultimate Theory.* New York: W.W. Norton, 1999. A best-selling but very challenging overview of recent development of "superstring theory," the branch of physics that attempts to unify all physical phenomena in a single description.

Gribbin, John, ed. *A Brief History of Science.* New York: Barnes & Noble Books, 1998.

Hawking, Stephen. *A Brief History of Time (The Updated and Expanded Tenth Anniversary Edition).* New York: Bantam Books, 1996. A best-selling book by the world's leading proponent of black-hole theories.

Hazen, Robert M., with Maxine Singer. *Why Aren't Black Holes*

Black? The Unanswered Questions at the Frontiers of Science. New York: Anchor Books, 1997.

Hazen, Robert M., and James Trefil. *Science Matters: Achieving Scientific Literacy*. New York: Doubleday, 1991. An overview of a variety of introductory science information.

Hellemans, Alexander, and Bryan Bunch. *The Timetables of Science*. New York: Simon and Schuster, 1987.

Hoskin, Michael, ed. *The Cambridge Illustrated History of Astronomy*. Cambridge, England: Cambridge University Press, 1997. A heavily illustrated, textbookish overview of advances from ancient times to modern astronomy.

James, Peter, and Nick Thorpe. *Ancient Mysteries*. New York: Ballantine Books, 1999. An engaging look at such mysteries as the Easter Island statues, druids, the Sphinx, and the pyramids.

Jastrow, Robert. *God and the Astronomers (New and Expanded Edition)*. New York: Norton, 1992. An examination of cosmology and religious traditions.

Jastrow, Robert. *Red Giants and White Dwarfs (New Edition)*. New York: Norton, 1990. A classic, now slightly dated, guide to the solar system and stars.

Koestler, Arthur. *The Sleepwalkers: A History of Man's Changing Vision of the Universe*. London: Penguin books, 1959. A provocative but somewhat dated history that includes the contributions of such scientists as Kepler and Galileo.

Krauss, Lawrence M. *The Physics of Star Trek*. New York: Basic Books, 1995. Can we teleport? What is actually possible in the famed television and film series?

Krupp, E.C. *Echoes of the Ancient Skies: The Astronomy of Lost Civilizations*. New York: Oxford, 1983. A somewhat dated and academic overview of astronomical views of the Greeks, Chinese, Babylonians, and others.

Leeming, David, and Margaret Leeming. *A Dictionary of Creation Myths*. New York: Oxford University Press, 1994.

Levy, David H., ed. *The Scientific American Book of the Cosmos*. New York: St. Martin's Press, 2000. A compilation of articles by writers from Albert Einstein to Carl Sagan, from the oldest and most popular science magazine in the world.

Light, Michael, and Andrew Chaikin. *Full Moon*. New York: Knopf, 1999. A visually spectacular photographic commemoration of the thirtieth anniversary of the *Apollo 11* lunar landing.

Maddox, John. *What Remains to Be Discovered: Mapping the Secrets of the Universe, the Origins of Life, and the Future of the Human Race*. New York: Free Press, 1998. A fascinating exploration of the major discoveries that still await science.

Moring, Gary F. *The Complete Idiot's Guide® to Understanding Einstein*. Indianapolis: Alpha Books, 2000. A highly approachable, readable guide to Einstein's thinking.

Odenwald, Sten. *The Astronomy Cafe: 365 Questions and Answers from* Ask the Astronomer. New York: W. H. Freeman, 1998. Drawn from the author's popular Web site.

Paterniti, Michael. *Driving Mr. Albert: A Trip Across America with Einstein's Brain*. New York: Dial Press, 2000. Amusing account of a cross-country car trip with Einstein's brain in a Tupperware bowl with the pathologist who autopsied the famed physicist.

Pirani, Felix, and Christine Roche. *Introducing the Universe*. Cambridge, England: Icon Books, 1993. A tongue-in-cheek overview of cosmology with cartoon-style illustrations.

Piszkiewicz, Dennis. *Wernher von Braun: The Man Who Sold the Moon*. Westport, Connecticut: Praeger, 1998. A damning account of the famed rocket scientist's Nazi past based upon recently released government documents.

Rankin, William. *Introducing Newton*. Cambridge, England: Icon Books, 1993. A cartoon-style overview of Newton's life and contributions.

Reston, James Jr. *Galileo: A Life*. New York: HarperCollins, 1994. A comprehensive biography of one of the greatest minds in science, his challenge to the Roman Catholic Church, and the resulting trial.

Sagan, Carl. *Cosmos*. New York: Random House, 1980. A best-selling companion to a PBS series that aired in 1980, written by one of the greatest proponents of space exploration.

Sobel, Dava. *Galileo's Daughter: A Historical Memoir of Science, Faith, and Love*. New York: Walker & Co., 1999. A best-selling account of the rich relationship between the Italian genius and his daughter who lived in a convent. Provides a rich biography of Galileo.

Sproul, Barbara C. *Primal Myths: Creation Myths Around the World.* New York: HarperCollins, 1991. An account of how different cultures have seen the origins of the universe.

Stott, Carole. *New Astronomer.* New York: DK Publishing, 1999. A practical guide to the skills and techniques of skywatching, aimed at younger readers.

Trefil, James. *101 Things You Don't Know About Science and No One Else Does Either.* Boston: Houghton Mifflin, 1996. Popular science writer's overview of the major challenges facing science in a variety of fields including astronomy and physics.

———. *Other Worlds: Images of the Cosmos from Earth and Space.* Washington, D.C.: National Geographic, 1999. Extraordinary photographic images of the universe in the brilliant tradition of the National Geographic Society.

Verdet, Jean-Pierre. *The Sky: Mystery, Magic and Myth.* New York: Abrams, 1987. A heavily illustrated overview of astronomy as it was reflected in art.

von Däniken, Erich. Translated by Michael Heron. *Chariots of the Gods: Unsolved Mysteries of the Past.* New York: Berkley Books, 1999. A sensational best-seller that began much of the excitement over the possibility of ancient visitors from other worlds.

Weinberg, Steven. *The First Three Minutes: A Modern View of the Origin of the Cosmos.* New York: Basic Books, 1993. An updated version of the classic explanation of the Big Bang theory by the winner of the 1979 Nobel Prize for Physics.

Williams, William F., ed. *Encyclopedia of Pseudoscience: From Alien Abductions to Zone Therapy.* New York: Facts on File, 2000. A listing of many popular beliefs with emphasis on refuting them according to evidence and science.

Wolfe, Tom. *The Right Stuff.* New York: Farrar, Straus & Giroux, 1979. The story of the astronauts and their families that NASA didn't want to tell you.

Wolke, Robert L. *What Einstein Told His Barber: More Scientific Answers to Everyday Questions.* New York: Dell Publishing, 2000. An amusing collection of all those oddities people wonder about, like what happens when you shoot a bullet in the air.

World Almanac and Book of Facts 2001. Cleveland: World Almanac Books, 2001.

Youngson, Robert. *Scientific Blunders: A Brief History of How Wrong Scientists Can Sometimes Be.* New York: Carroll & Graf, 1998. An entertaining overview of some of the embarrassments of scientific history.

PERIODICALS

Ad Astra. The publication of the National Space Society.

Astronomy. A monthly aimed at serious amateur astronomers.

Discover. A lively, general-interest science magazine, which includes frequent astronomical features and space-related news stories alongside Bob Berman's "Sky Lights" column for skywatchers.

National Geographic. The classic photojournalist magazine frequently includes items about space.

Natural History. The magazine of the American Museum of Natural History and its affiliated Rose Science Center Planetarium. For the general-interest reader, it includes articles on various aspects of astronomy and the universe.

Planetary Report. The bimonthly magazine of the Planetary Society, an organization devoted to space exploration.

Scientific American. The oldest science magazine and still one of the best, although the level of writing is often very challenging.

Sky & Telescope. Another excellent magazine for amateur astronomers.

WEB SITES

The Internet can be wonderful and confounding. A search of America OnLine yields more than 64,000 sites under "space" and another 8,000 under "astronomy." The following abbreviated list includes authoritative sites with valid links to many other sites, including direct links to pages of spacecraft such as the International Space Station, the Hubble Space Telescope, Chandra X-Ray Observatory, and many other spacecraft currently transmitting data.

astronomy.com   *Astronomy* magazine online, the most popular sky-watching magazine.

badastronomy.com   Run by astronomer Phil Plaitt, a site dedicated to overcoming astronomy myths and public confusion.

iau.org   Official site of the International Astronomical Union.

nasa.gov   Home site of the National Air and Space Administration.

sohowww.nascom.nasa.gov   A NASA site dedicated to the SOHO satellite studying the Sun.

neat.jpl.nasa.gov   Official NASA site that tracks asteroids and other near-Earth objects.

nationalgeographic.com   Home page of the famed magazine online, with photographs and articles on space.

planetary.org   Home page of the Planetary Society, a private group dedicated to continued exploration of space.

sciam.com   Home page of the *Scientific American* magazine online.

space.com   A commercial site, meaning banner advertising, but includes excellent links to many other sites.

seti-inst.edu   Home site of the Search for Extraterrestrial Intelligence.

lpl.arizona.edu/spacewatch   A University of Arizona site that keeps track of asteroids

nss.org   Home page of the National Space Society.

nss.org/adastra/home   The National Space Society and its magazine *Ad Astra*.

fly.hiwaay.net/~hal5/space-links.shtml   More than five thousand valid space-related links maintained by a chapter of the National Space Society.

planetary.org   Home page of the Planetary Organization.

seds.org   Site maintained by the Students for Exploration and Development of Space, hosted by the University of Arizona chapter.

sci.esa.int/index.cfm   Home site of the European Space Agency, with links to the *Hipparchos* satellite.

whyfiles.org   University of Wisconsin site that provides answers to questions about science, including astronomy.

yahoo.com/science/astronomy   A Yahoo directory of links to science and astronomy sites.

INDEX

About the Author

KENNETH C. DAVIS, recently dubbed the "King of Knowing," speaks often on national television and radio. His *USA Weekend* column is read weekly by millions. In addition to his adult titles, he writes the Don't Know Much About® children's series, published by HarperCollins. Davis is a contributing editor to *USA Weekend* magazine, where his weekly quizzes are read by millions. He lives in New York City with his wife and two children.